Troubleshooting for Network Operators

To my sweet love Ly Ly and beloved family
Van Van Tong

To all my family, my wife, my dear children Sarah and Sinan, with all my love and infinite gratitude for your presence in my life
Sami Souihi

To my dearest family, thank you for being my constant source of love and support throughout my life
Hai-Anh Tran

To my beloved, sublime and thoughtful daughter, Kenza Insafe, on her 20th birthday this year
Abdelhamid Mellouk

New Generation Networks Set

coordinated by
Abdelhamid Mellouk

Volume 3

Troubleshooting for Network Operators

The Road to a New Paradigm with Encrypted Traffic

Van Van Tong
Sami Souihi
Hai-Anh Tran
Abdelhamid Mellouk

iSTE

WILEY

First published 2023 in Great Britain and the United States by ISTE Ltd and John Wiley & Sons, Inc.

ISTE Ltd
27-37 St George's Road
London SW19 4EU
UK

www.iste.co.uk

John Wiley & Sons, Inc.
111 River Street
Hoboken, NJ 07030
USA

www.wiley.com

Library of Congress Control Number: 2023938907

British Library Cataloguing-in-Publication Data
A CIP record for this book is available from the British Library
ISBN 978-1-78630-867-2

Contents

Preface ix

Introduction xi

Chapter 1. State of the Art on Network Troubleshooting 1

1.1. Network troubleshooting 1
1.1.1. State of the art 2
1.1.2. Traditional troubleshooting architecture 9
1.2. Background on encryption protocols 10
1.2.1. QUIC 11
1.2.2. Other protocols 16
1.3. Drawbacks of troubleshooting with encrypted traffic 18
1.3.1. Network performance monitoring 18
1.3.2. Intrusion detection system 20
1.4. Conclusion 22

Chapter 2. Novel Global Troubleshooting Framework for Encrypted Traffic 25

2.1. Novel network troubleshooting architecture for encrypted traffic 25
2.2. Proof of concept of novel troubleshooting architecture in SDN 28
2.3. Data collection 32
2.3.1. Data classification 32
2.3.2. Monitoring tools 34

2.3.3. Parameter measurement 37
2.4. Troubleshooting dataset 40
2.4.1. Datasets for root cause analysis 40
2.4.2. Dataset for traffic classification 42
2.5. Conclusion 43

Chapter 3. Traffic Classification: Novel QUIC Traffic Classifier Based on Convolutional Neural Network 45

3.1. Introduction 45
3.2. Background 48
3.2.1. Convolutional network 48
3.2.2. Characteristics of QUIC-based applications 49
3.3. Traffic classification approaches 50
3.3.1. Port-based approaches 50
3.3.2. Payload-based approaches 51
3.3.3. Statistic-based approaches 51
3.3.4. DL-based approaches 52
3.4. Novel traffic classification method for QUIC traffic 53
3.4.1. Traffic collection 55
3.4.2. Flow-based features 55
3.4.3. Preprocessing 56
3.4.4. Novel traffic classification method 56
3.5. Experimental results 59
3.5.1. Dataset specification 59
3.5.2. Performance metrics 60
3.5.3. Performance analysis 61
3.6. Conclusion 65

Chapter 4. Anomaly Detection 67

4.1. Introduction 67
4.2. Anomaly detection approaches 68
4.2.1. Knowledge-based mechanisms 68
4.2.2. Rule inductions 69
4.2.3. Information theory 70
4.2.4. ML-based mechanisms 70
4.3. Anomaly detection approach using machine learning 71
4.3.1. ML-based anomaly detection method 72
4.3.2. Data collection and processing 74

4.4. Experimental results 75
4.4.1. Experimental setup 75
4.4.2. Performance analysis 76
4.5. Conclusion 79

Chapter 5. Temporary Remediation: SDN-based Application-aware Segment Routing for Large-scale Networks 81

5.1. Introduction 81
5.2. Application-aware routing mechanisms 84
5.2.1. Application-aware routing 84
5.2.2. Application-aware MPLS 86
5.2.3. Application-aware SR 86
5.3. Adaptive segment routing mechanism for encrypted traffic . . . 87
5.3.1. Overview of the SDN-based adaptive segment routing framework 87
5.3.2. Network monitoring 89
5.3.3. Anomaly detection 90
5.3.4. Application-aware remediation 91
5.4. Experimental results 95
5.4.1. Experiment setup 95
5.4.2. Benchmark 97
5.4.3. Performance analysis 97
5.5. Conclusion 104

Chapter 6. Root Cause Analysis and Definitive Remediation . 107

6.1. Root cause analysis: machine learning based root cause analysis for SDN network 107
6.1.1. Introduction 107
6.1.2. Root cause analysis mechanisms 109
6.1.3. ML-based RCA mechanism 111
6.1.4. Experimental results 114
6.1.5. Conclusion 119
6.2. Definitive remediation: adaptive QUIC BBR algorithm using reinforcement learning for dynamic networks 121
6.2.1. Introduction 121
6.2.2. Congestion control mechanisms 123
6.2.3. Adaptive BBR algorithm 126

6.2.4. Experimental results 128
6.2.5. Conclusion 133

Conclusions and Prospects 135

References 141

Index 159

Preface

The Internet network is continuously growing in terms of size and usage. This expansion results in an increase in complexity emphasized by a prevalence of encrypted traffic. This leads to complex problems in the network, bringing many negative economic impacts for network operators. In the past, network troubleshooting solutions were extensively studied to troubleshoot network problems and deal with them completely. However, conventional solutions are unmanageable because of high time-consuming, inadaptability for encrypted traffic, scalability, overhead, automation, and efficiency.

In this book, we provide insight into network troubleshooting in the context of encrypted traffic, and present a solution to these problems, which consists of an adaptive mechanism. The objective is to detect network anomalies, diagnose their root causes, and address them definitively. It highlights the necessity of network troubleshooting and economic impacts on network operators. Additionally, it provides a state of the art on network problems (e.g. rule failure, link failure and so on) as well as existing approaches. Moreover, the book thoroughly analyzes encryption protocols before analyzing the limitations of conventional troubleshooting solutions. Furthermore, it contains a novel troubleshooting architecture in the context of encrypted traffic and presents a proof of concept of this architecture in the software-defined networking (SDN) environment. This architecture contains five main modules: data collection, anomaly detection, temporary remediation, root cause analysis and definitive remediation.

Giving a complete bibliography and a historical account of the research that led to the present form of the subject would be impossible. Thus, some topics are discussed in less detail than others. The choices made reflect, in part, personal taste and expertise and, in part, a preference for up-and-coming research and recent developments in network troubleshooting for network operators.

This book is a preliminary for network troubleshooting in the context of encrypted traffic. We hope that it will inspire other researchers and network operators on the road of network troubleshooting. The authors hope you will enjoy reading this book and get many helpful ideas and overviews for your own studies.

Van VAN TONG
Sami SOUIHI
Hai ANH TRAN
Abdelhamid MELLOUK
June 2023

Introduction

> *"We are all now connected by the Internet, like neurons in a giant brain"*
>
> Stephen Hawking

Network troubleshooting plays an important part in the network. This is because computer networks have rapidly evolved along with the significant growth of the Internet of Things (IoT), increasing not only a network coverage, but also the complexity in the computer network, bringing a risk of incurring problems in the network. For example, there are many problems in the network, including server disruptions, cyberattacks, link failure, etc. Therefore, network troubleshooting, which is a process of detecting an anomaly, identifying its root causes and implementing remediation approaches to solve it definitively, is studied thoroughly by the research community (see Fonseca and Mota (2017); Yu et al. (2018); Cherrared et al. (2019)).

Regarding server disruptions, Table I.1 illustrates the total downtime and corresponding money lost for several service providers (Gagnaire et al. 2012). For instance, YouTube and PayPal lost from $34,000 to over $6,700,000 because of a few hours of disruption on their cloud servers due to failures.

Besides, many cloud services nowadays are disrupted by cyberattacks (e.g. distributed denial of service (DDoS) attacks, etc.). A DDoS attack is a kind of cyberattack designed to overload and disrupt network services by exhausting them with access requests. In February 2020, Amazon announced that their AWS Shield services mitigated the largest recorded DDoS attack with 2.3 Tbps of network traffic (Felter 2021). This attack, which caused three days of "elevated threat" for AWS Shield services, was carried out using hijacked Connection-less Lightweight Directory Access Protocol (CLDAP) web servers.

	Total downtime (h)	Cost (USD)
YouTube	0.17	34,000
CloudFlare	1	168,000
Zoho	33.5	600,000
Cisco	5.33	1,066,000
eBay	6.25	1,406,250
Facebook	8.5	1,700,000
PayPal	30.2	6,795,000

Table I.1. *Downtime of service providers and their economic impacts*

To deal with network problems, network troubleshooting has been extensively studied for over 30 years. However, conventional solutions are not effective because of high time consumption, inadaptability for encrypted traffic, overheads, scalability, etc.

Regarding the high time consumption, the processing time for root cause analysis and remediation can take from 1 h to more than 5 h depending on the status of anomalies in the network (Zeng et al. 2012b). As a result, network systems can suffer from negative impacts (e.g. high latency, high loss, etc). These impacts can result in frequent connection interruptions in the network. Depending on the anomaly's nature, there are two possible cases. If the root cause of the anomaly is identified and solved quickly, temporary remediation is not necessary. Otherwise, temporary remediation is required to guarantee the availability of the network. Therefore, it became inevitable to think about network troubleshooting frameworks that guarantee the network's availability during the root cause analysis and definitive remediation.

As for the indaptability, the traditional troubleshooting mechanisms were not designed for encrypted traffic. However, many service providers today encrypt network traffic to prevent attackers from inspecting data packets for illegal activities. Concretely, 80% of web traffic was encrypted by 2019 compared to 40% by 2016 (Cisco 2021a). From the point of view of network operators (NOs), the information in the packets is hidden, such as the sequence number, acknowledgment number, payload signatures, etc. This introduces several limitations related to network performance monitoring approaches (e.g. estimation of quality of experience, application identification, etc.) and intrusion detection systems (Kühlewind et al. 2018; Moriarty and Morton 2018). Therefore, encrypted traffic results in many obstacles for troubleshooting, particularly in data collection (e.g. collecting performance metrics, etc.) and remediation approaches using a deep packet inspection (e.g. application-aware traffic engineering, signature-based intrusion detection systems, etc.).

Concerning the overhead, carrying out data collection without influencing the network performance is a challenging task. In fact, the continuous monitoring of network data and network traffic can generate a huge overhead on the network and thus influence the network performance. Many network monitoring proposals have developed different approaches to balance data accuracy and monitoring overhead.

As for the scalability, in traditional network architectures, control logic is distributed in network devices, so updated policies in network troubleshooting are implemented separately in each network device. Nowadays, the number of network devices increases rapidly due to the rapid growth of the Internet network and IoT. This leads to the scalability issue for network troubleshooting, as well as network management.

In this book, we put a special focus on the high time consumption and inadaptability issues. Concretely, we present an analysis and comments on the network troubleshooting mechanisms before elaborating our vision of a troubleshooting framework for network operators in the context of encrypted traffic. This framework is composed of five modules: data collection, anomaly detection, temporary remediation, root cause analysis and definitive remediation. In addition to the troubleshooting framework, this book focuses on four essential points:

– in addition to the traditional data collection module, it is necessary to think about a novel traffic classification approach to classify encrypted traffic into different kinds of applications (e.g. video streaming, file transfer, etc.). In fact, the application class plays an important role in the remediation approaches (e.g. application-aware mechanisms, etc.) in network troubleshooting;

– a temporary remediation approach to assure the availability of the network as well as meet strict SLA requirements during the root cause analysis and definitive remediation;

– a proof of concept for the root cause analysis and definitive remediation in network troubleshooting that allows us to automatically identify the root cause of anomalies and address it completely;

– troubleshooting datasets: we build and contribute the troubleshooting datasets that contain a dataset for encrypted traffic classification approaches and two datasets for the root cause analysis, in order to facilitate the network troubleshooting.

In this book, each chapter is dedicated to one module of the troubleshooting framework. The remainder of the book is organized as follows:

Chapter 1: provides related work on several network problems (e.g. link failure, switch failure, etc.). In addition to explaining the fundamental parts of traditional troubleshooting architecture, we explain how network traffic is encrypted and provide an analysis on the limitations of network troubleshooting for encrypted traffic.

Chapter 2: presents fundamental parts of novel troubleshooting architecture in the context of encrypted traffic and shows a proof of concept of this architecture in a software-defined networking (SDN) environment. We also present a parameter measurement module to collect data in order to build troubleshooting datasets. Moreover, the chapter thoroughly describes the troubleshooting datasets, which are composed of datasets for root cause analysis and encrypted traffic classification.

Chapter 3: describes a novel encrypted traffic classification method to identify different kinds of applications. The purpose is to provide information about application classes for application-aware mechanisms in network troubleshooting.

Chapter 4: presents related work on anomaly detection. Moreover, this chapter takes into account congestion to generate anomalies and presents an anomaly detection approach using machine learning to detect these anomalies in the network.

Chapter 5: presents an application-aware segment routing mechanism in temporary remediation. This mechanism identifies application classes according to traffic classification. In a particular application, this mechanism implements a specific routing strategy based on a reinforcement learning algorithm to meet strict SLA requirements.

Chapter 6: considers congestion as a use-case for the root cause analysis and definite remediation. This chapter presents a root cause analysis using machine learning to identify the root cause of congestion. It also presents an adaptive congestion control algorithm to solve it completely.

Conclusions and Prospects: this last chapter concludes this book and provides an insight into the future work and prospects in the area of network troubleshooting.

1

State of the Art on Network Troubleshooting

"A protocol approach to troubleshooting"

Ed Wilson

Chapter 1 presents the state of the art on network troubleshooting and a traditional troubleshooting architecture for non-encrypted traffic. We then discuss its limitations when traffic is encrypted.

1.1. Network troubleshooting

In the early 19th century, technicians were dispatched to find problems in telegraph and phone line infrastructure to repair and solve the issues. Historically, a troubleshooter refers to a skilled worker who finds and solves technical problems. Nowadays, *troubleshooting* is a form of problem-solving that aims to repair failed processes in a machine or a system. According to the related work Morris and Rouse (1985) and Jonassen and Hung (2006), there are several existing conceptions of the troubleshooting process. The basic concept of troubleshooting is finding the faulty components in a device to repair or replace it Perez (1991). Schaafstal et al. (2000) designed the troubleshooting process with four subtasks: formulating problem description, cause generation, test and evaluation. Similarly, troubleshooting is considered as an iterative process with four subprocesses: problem space construction, problem space reduction, fault diagnosis and solution verification (Johnson et al. 1993).

Network troubleshooting is an iterative process with three subtasks: identifying, diagnosing and solving problems in the network. In the past, network operators (NOs) implemented manual troubleshooting tools such as *ping*, *traceroute*, etc. *ping* is a computer network administration utility designed to check a reachability between a source and a destination and round-trip time of packets in the network. *traceroute* is a computer network diagnostic utility used to display possible routes between a source and a destination and measure a transit delay of packets in the network. These troubleshooting tools are used to diagnose complex problems such as loops caused by undefined interaction between spanning tree protocols (Heller et al. 2013), etc. However, these approaches are not effective with a huge number of network devices. Besides, 24.6% of administrators reported that anomaly diagnosis takes more than 1 h on average to solve anomalies (Zeng et al. 2012a). Therefore, it is necessary for an automated troubleshooting process that aims to detect an anomaly, locate its causes and solve it. Consequently, network troubleshooting is considered by the research community Fonseca and Mota (2017); Yu et al. (2018); Cherrared et al. (2019). In the following section, we present the state of the art of network troubleshooting.

1.1.1. *State of the art*

According to the related work on network troubleshooting (Yu et al. 2018; Fonseca and Mota 2017; Van et al. 2018), problems can be classified into several categories thanks to locations where problems happen or factors that result in problems. Yu et al. (2018) and Fonseca and Mota (2017) categorize problems into problems in application, control and infrastructure layer. Similarly, problems can be classified into problems in application service providers (ASP) or Internet service providers (ISP) (Van et al. 2018). Besides, problems can be classified into problems caused by administrators (e.g. router misconfiguration, server misconfiguration, etc.) or problems that are not caused by administrators (e.g. link failure, switch failure, buffer overload, etc.). According to a survey of NOs (Zeng et al. 2012b), in this book, we present several problems that are not caused by administrators in following sections.

1.1.1.1. *Rule failure*

Bu et al. (2016) categorized failure rule in the network into missing fault and priority fault. The missing fault occurs when a rule is not executed as expected, whereas the priority fault occurs when overlapping rules violate a priority order.

There are research studies concentrating on the missing fault including ATPG in Zeng et al. (2012a) and Monocle Perešíni et al. (2015). These approaches verify the rules by generating probe packets to exercise every rule. ATPG uses a header space analysis (Kazemian et al. 2012) to check the reachability between all test hosts. Then, the reachability result is transferred to a probe packet generator to compute a minimal set of probe packets via greedy algorithm (Slavık 1997). Next, these probe packets are sent into the network systems to check the rule's corrections. If an error is detected, a fault localization algorithm is implemented to narrow down to identify the root cause. However, ATPG has a drawback when it generates the probe packets for all rules. It is not effective when there are only a few up-to-date rules. Consequently, Monocle is proposed to overcome this drawback. This approach only verifies recently installed rules and reports misbehaviors. Besides, Monocle formulates knowledge from flow tables in the switches as constraints and applies an SAT solver (Biere 2008) to generate a set of probe packets.

Probing is an intrusive method that generates significant overheads and increases link utilization in the network. Consequently, it is necessary to minimize the number of probe packets. This is a minimum set cover problem, which is an NP-Complete problem (Zeng et al. 2012a). Therefore, Bu et al. (2016) proposed RuleScope, a framework for detecting rule failures in the network. RuleScope divides flow tables into solvable subsets of rules to minimize probe scale. Then, this approach creates a directed acyclic graph for each subset and generates a set of probe packets for each subset. As a result, this approach processes the probe packet generation more quickly due to a small scale of rule subsets.

Although RuleScope minimizes the number of probe packets, this approach suffers from a drawback related to a separation in the flow tables. This leads to the priority fault in the switches. The separation in the flow tables into small subsets can result in pretermitting two overlapping rules in two different subsets of rules. Zhao et al. (2018a) proposed SERVE, a rule verification to identify rule failure in the switches automatically. Firstly, SERVE extracts all rules for each device and builds a multi-rooted tree that considers rule connections. Next, SERVE analyzes the multi-rooted tree to generate the minimum number of probe packets. The minimum set cover problem is an NP-Complete problem, so SERVE applies the depth-first search (DFS) algorithm to generate the probe packets. Zhao et al. (2018b) extended the previous study of Zhao et al. (2018a) to present a complete framework. After generating the probe packets, SERVE injects these packets into network systems using an out-band channel. Besides, SERVE also computes a desired network behavior using the multi-rooted trees. According to a comparison between the feedback from the out-band channel for every rule and the desired network behaviors, SERVE can detect faulty rules and send notifications to administrators. SERVE's performance is evaluated to benchmarks in processing time, number of probe packets and overheads. Concerning the number of probe packets, SERVE decreases the number of probe packets by up to 75% in comparison with Monocle. Regarding the processing time, SERVE's figure is three times less than the figure for ATPG. As for the overhead, in-band bandwidth is not influenced according to using the out-band channel to inject the probe packets. Besides, the out-band bandwidth is far less than link capacity.

1.1.1.2. *Link failure*

Link failure refers to unreachability between two switches. It can lead to a high packet loss and performance degradation in the network. Link failure can be detected according to probe packets in active monitoring approaches. *ping* is a simple troubleshooting tool that sends probe packets to check the reachability between two end-points. If probe packets are lost, it means that there is a faulty link between these end-points. Similarly, Cascone et al. (2017) proposed a fast failure detection mechanism to detect the link failure based on the exchange of bidirectional "heartbeat" packets. When the packet rate drops below a threshold, a node sends heartbeat packets to its neighbors. If there are no responses from its neighbors after a given time, the link failure happens in the network. However, this mechanism requires a strict consumption related to the backup solutions that cannot be utilized to guarantee the short failover delays (1 ms).

Moreover, this problem can be detected by using the Link Layer Discovery Protocol (LLDP) in software-defined networking (SDN) (Khan et al. 2016; Tarnaras et al. 2015). According to the topology discovery protocol, SDN controller can detect link failure and remove it from network topology. Firstly, an OpenFlow (OF) switch connects to the controller so that the controller knows its active ports. Next, the controller generates a Packet-out message to each active port in the switch to discover the topology. The LLDP between switch s_1 and s_2 is depicted in Figure 1.1. Firstly, the controller encapsulates an LLDP packet in a Packet-out message and sends it to the switch s_1. When switch s_1 receives the Packet-out message, it will forward the LLDP packet to switch s_2. After receiving the LLDP packet, switch s_2 encapsulates this packet in a Packet-in message and sends it back to the controller. The controller receives this message and creates a link from switch s_1 to s_2. The same process is performed to identify the link for an opposite direction. When link s_1–s_2 is faulty, the controller will not receive the Packet-in message from switch s_2. Then, the controller will remove this link from the network topology. In the network with S switches interconnected by a set of L links, the total number of Packet-out and Packet-in messages are described in equations [1.1] and [1.2], respectively. P_i is the number of the active port in the switch S_i.

$$TOTAL_{PACKET-OUT} = \sum_{i=1}^{S} P_i \quad [1.1]$$

$$TOTAL_{PACKET-IN} = 2L \quad [1.2]$$

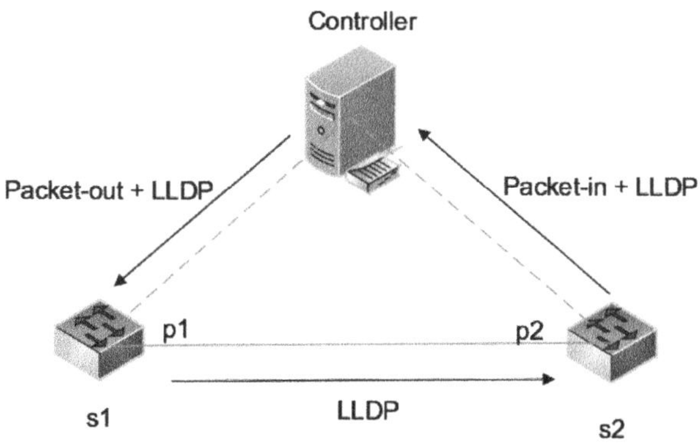

Figure 1.1. *Unidirectional link discovery in LLDP. For a color version of this figure, see www.iste.co.uk/tong/troubleshooting.zip*

Unlike the SDN environments, a hybrid SDN contains OF switches and traditional switches that LLDP cannot discover. Therefore, SDN controllers (e.g. Floodlight Floodlight (n.d.), OpenDayLight ?, etc.) utilize a combination of LLDP and Broadcast Domain Discovery Protocol (BDDP) (Khan et al. 2016) for the topology discovery. There are two main differences between LLDP and BDDP. First of all, destination MAC address in BDDP is a broadcast address (FF:FF:FF:FF:FF:FF) in contrast to multicast addresses used by LLDP. This allows the traditional switch forwarding the BDDP packets to detect multi-hop links, whereas LLDP is a mechanism for a single-hop link detection. Secondly, EtherType of LDDP is 0x88cc while EtherType of BDDP is 0x8999.

To keep up-to-date topology information, the controller needs to send $TOTAL_{PACKET-OUT}$ Packet-out messages and $TOTAL_{PACKET-IN}$ Packet-in messages periodically. This leads to a significant control overhead and an increase in link utilization. Therefore, many studies focus on optimizing the number of control packets for the topology discovery. Pakzad et al. (2014) proposed OpenFlow Discovery Protocol version 2 (OFDPv2) to reduce the number of Packet-out messages. Firstly, OFDPv2 reduces the number of LLDP Packet-out messages sent to each switch to one. Then, this protocol installs a new rule in each switch to forward LLDP packets received from the controller to all available ports. Finally, OFDPv2 customizes the event handler of the Packet-in message in the controller to parse the MAC address.

Besides, Xu et al. (2017) proposed a monitoring link failure detection approach to reduce the number of Packet-out messages in LLDP to one. Firstly, this approach calculates a monitoring tree path based on network topology. Then, this approach installs monitoring flow entries into each switch based on the switch's locations in the monitoring tree path to match and apply actions to monitoring packets. Next, the controller sends a Packet-out message to a source node to instruct this node to send out the monitoring packets to other nodes following the monitoring tree. If a switch cannot receive the LLDP packet after a given time, this switch will send an alarm packet to the controller.

Unlike sending LLDP packets periodically to network systems to update the topology information, Azzouni et al. (2018) proposed sOTDP, a secure and efficient OpenFlow Discovery Protocol that applies bidirectional forwarding detection (BFD) (Katz and Ward 2010) to detect link's events quickly. The controller listens to link's event notifications from switches to update the network topology instead of updating it periodically by sending LLDP packets. BFD is a detection protocol designed to provide a fast failure detection of forwarding paths. This protocol establishes a session between two endpoints over a particular link and sends control messages to detect active links.

1.1.1.3. *Buffer overload*

Switch's buffer overload occurs when a data volume exceeds the storage capacity of buffer memory in switches. This can lead to congestion in the network. Phanishayee et al. (2008) highlighted that an increase in buffer size can solve the buffer overload due to a massive amount of network traffic. Nevertheless, this approach has low scalability because the replacement cost of conventional devices is extremely high. Therefore, many studies focus on congestion control and congestion avoidance mechanisms to prevent and avoid the buffer overload.

Concerning the congestion control mechanisms, there are two kinds of flows in the network: elephant flows and mice flows. Elephant flows are the large continuous flows that consume high bandwidth, whereas mice flows are the small flows that consume low bandwidth. The elephant flows contain more than 90% of all bytes transmitted in the network (Kandula et al. 2009; Benson et al. 2010). Thus, Kanagevlu and Aung (2015) proposed a local rerouting mechanism for elephant flows in the SDN environment to decrease negative impacts of buffer overload. If link utilization is higher than 75% of link capacity, an alarm is triggered. The controller reroutes the elephant flows into other paths with minimal link utilization to avoid the congested links. As for the mice flows, its lifetime is short, but the percentage of mice flows in network traffic is very high, approximately 90% (Kandula et al. 2009; Benson et al. 2010). Trestian et al. (2013) proposed MiceTrap, an OpenFlow-based traffic engineering approach for mice flows. Firstly, the network traffic is classified into elephant flows and mice flows. The elephant flows are processed as in previous research studies (e.g. Kanagevlu and Aung (2015), etc.), whereas the mice flows are processed by MiceTrap. If a traffic volume exceeds a threshold, MiceTrap is triggered. Then, MiceTrap routes the mice flows using a weighted multi-path routing algorithm, which assigns these flows into paths based on its weight. The higher the path's weight, the less the path is chosen.

Besides, there are studies focusing on both elephant and mice flows in the routing approach. Song et al. (2016) proposed a routing algorithm which reroutes congested links when link utilization is higher than 70% of link capacity. However, the routing algorithms without considering available bandwidth can lead to congested links in the network. Thus, Attarha et al. (2017) and Gholami and Akbari (2015) proposed a routing algorithm that estimates link utilization and selects the links with high available bandwidth. Similarly, Sminesh et al. (2018) proposed a routing algorithm using a Bayesian network. This approach contains three modules. The first one is to identify a bottleneck link when its link utilization is higher than a threshold. The second one is to update topology and identify all available routing paths. The final one is to select the routing paths using the Bayesian network according to link utilization and residual bandwidth.

The above routing mechanisms consider a routing policy for different applications (e.g. video streaming, file transfer, etc.). However, it is ineffective because each application contains specific SLA requirements (e.g. low latency, high bandwidth, etc.). Consequently, many research work placed a special focus on application-aware routing algorithms. Li et al. (2016) proposed an application-aware traffic control scheme that implements a simple flow classification to apply corresponding rules. Adami et al. (2015) proposed a differentiated routing algorithm in the network. Firstly, this approach applies a deep packet inspection to identify the Session Initiation Protocol (SIP) flows using its signatures (e.g. UDP packet with destination port 5060, content-Type is set to application/sdp, etc.). Next, this approach calculates the shortest paths for these flows using the Dijkstra algorithm with link utilization as a link cost. Similarly, Cheng et al. (2015) proposed an application-aware routing algorithm to implement different routing policies corresponding to various applications. Firstly, this method classifies the flows into three categories consisting of real-time applications (VoIP, video communication and gaming), streaming applications (streaming video, audio, IPTV and web browsing) and miscellaneous applications (file sharing and miscellaneous upload/download). Then, each category is processed with a specific routing policy using different network parameters such as link load, delay and delay variation. Thomas et al. (2016) proposed AMPF, an application-aware multi-path packet forwarding framework in SDN using a machine learning algorithm. First, AMPF classifies network flows into real-time traffic, buffered transfer, web browsing and restricted transfer. Next,

AMPF calculates k available paths using Dijkstra's algorithm corresponding to each application. Finally, AMPF assigns network flows into these routing paths.

Regarding the congestion avoidance algorithms, Abdelmoniem and Bensaou (2015) proposed a switch-assisted TCP congestion control approach using a small modification to switches. When congestion is detected, this approach rewrites the receiver window field in TCP header to reduce sending rate. Similarly, Hwang et al. (2015) proposed SCCP, a scalable congestion control protocol. SCCP limits sending rate of TCP senders by leveraging switches. Concretely, SCCP modifies the advertisement window field in the TCP header of packets traversing the switches. Besides, many studies are focusing on end-to-end congestion avoidance algorithms which adjust sending rate at the sender's side (Zhang et al. 2019; Zhang and Mao 2020). Xu et al. (2019) proposed DRL-CC, a congestion control algorithm using deep reinforcement learning (DRL) in MPTCP. DRL-CC monitors acknowledgment packets to extract network parameters (e.g. RTT, RTT deviation, goodput, etc.) and implements a DRL algorithm to adjust the congestion window (cwnd) at the sender's side. Similarly, Jay et al. (2019) proposed Auro, an Internet congestion control algorithm using DRL. Auro monitors network states and adjusts sending rate using the DRL algorithm and network parameters including latency gradient, latency ratio and sending ratio.

1.1.2. *Traditional troubleshooting architecture*

Following existing survey on network troubleshooting (Fonseca and Mota 2017; Yu et al. 2018; Cherrared et al. 2019), the traditional troubleshooting architecture contains four main modules including data collection, anomaly detection, root cause analysis and remediation modules. The traditional troubleshooting architecture is depicted as in Figure 1.2. Firstly, the data collection module monitors the network systems and collects the network traffic using monitoring approaches (e.g. NetFlow (Suárez-Varela and Barlet-Ros 2017), sFlow (Ujjan et al. 2020), OpenFlow-based monitoring approaches, etc.) to extract troubleshooting data (e.g. network parameters, etc.). Then, these data are analyzed in the anomaly detection module to detect anomalies (moments when network problems happen) in the network. After that, the root cause analysis analyzes thoroughly the troubleshooting data to identify the root cause of anomalies in the network. An anomaly can be caused by many root causes. For instance, congestion can result from link failure, DDoS attack, switch's buffer overload, etc. Finally, root cause is solved definitively in the remediation module to return normal states to the network. The purpose of the troubleshooting architecture is to build self-healing networks that automatically detect, diagnose and remediate the network anomalies.

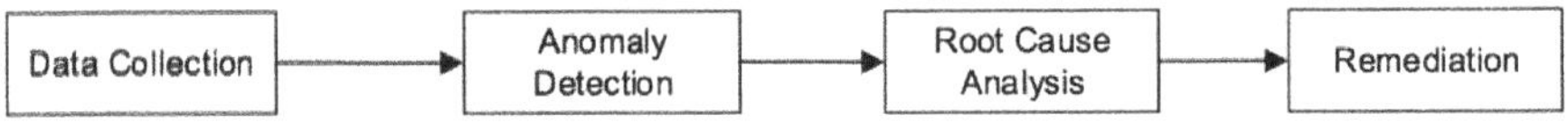

Figure 1.2. *Overall traditional troubleshooting architecture*

Facing a growing number of data stealing attacks, encrypted traffic is utilized by many service providers to make their data more secure and protect their user's privacy. This leads to many difficulties for network troubleshooting (Kühlewind et al. 2018). In the following sections, we present how network traffic is encrypted and drawbacks of network troubleshooting in the context of encrypted traffic.

1.2. Background on encryption protocols

In the report of Thales e-Security in April 2017 e-Security (n.d.), the percentage of encrypted flows ranges from 16% to 41% during the period 2005–2016. Figure 1.3 shows the global growth of encrypted traffic from 2005 to 2016. Similarly, in a current Cisco report Cisco (2021a), 80% of web traffic is encrypted by 2019 in comparison with 40% in 2016.

According to the related work of Velan et al. (2015), there are two kinds of encryption protocols in the network. The first one encrypts the whole packet (e.g. IPsec, etc.) while the second one only encrypts packets' payload (e.g. TLS, etc.). Besides, Google has recently developed Quick UDP Internet Connections (QUIC), a new transport layer network protocol from 2012, to provide secure connections and improve latency in the network (Langley et al. 2017). Similar to Transport Layer Security (TLS), QUIC only encrypts the payload and a part of packet's header. QUIC is equivalent to a combination of TCP and TLS. Moreover, QUIC offers opportunities to overcome the limitations of TCP + TLS related to connection establishment time, TCP Head of Line blocking and connection migration (Langley et al. 2017). Therefore, in this book, we consider QUIC as an encryption protocol in the network troubleshooting framework. The background about IPsec, TLS and QUIC are presented as follows.

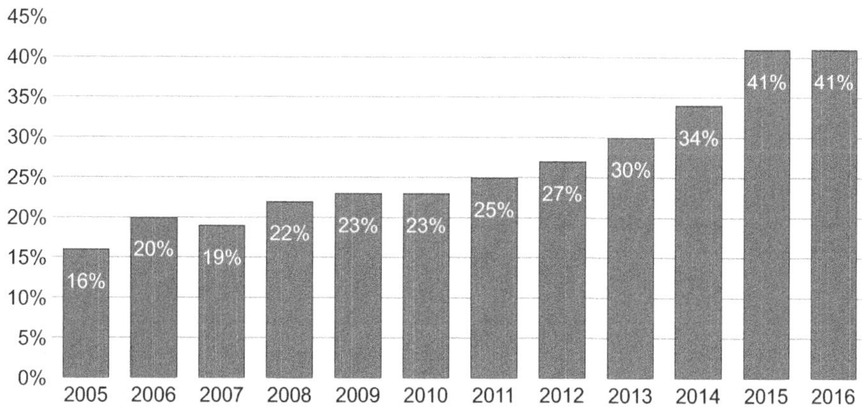

Figure 1.3. *The global growth of the encrypted traffic e-Security (n.d.). For a color version of this figure, see www.iste.co.uk/tong/troubleshooting.zip*

1.2.1. *QUIC*

QUIC (Langley et al. 2017) is a novel transport protocol designed by Google from 2012. Its design is motivated by multiple existing protocols such as TCP, TLS and HTTP/2 (Hypertext Transfer Protocol version 2). In early 2021, QUIC was standardized by Internet Engineering Task Force (IETF) to solve urgent issues of TCP + TLS (Kosek et al. 2021). The difference between TCP + TLS and QUIC is depicted as in Figure 1.4. QUIC is designed to improve latency in the network by sending data directly with 0-RTT in the best case. Moreover, it incorporates TLS1.3 to require all connections to be encrypted. QUIC also offers multiplexing features to send multiple data streams over a single connection. Furthermore, it offers opportunities to migrate connections without interrupting data communication. QUIC is implemented on the top of User Datagram Protocol (UDP), so it can be deployed in user space in contrast to TCP, which is implemented in system kernels. This feature allows us to deploy QUIC quickly in an application update cycle. The deep analysis on QUIC's characteristics (Cui et al. 2017) is described as follows.

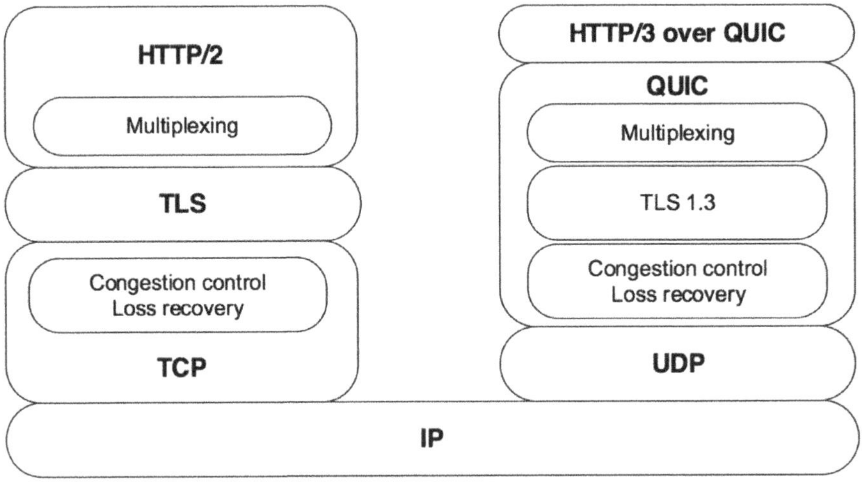

Figure 1.4. *Difference between TCP + TLS and QUIC architecture*

1.2.1.1. *Encryption*

QUIC uses TLS1.3 to provide end-to-end encryption. When the TLS handshake is complete, the encryption is performed on the UDP payload. In QUIC protocol, there are two kinds of headers: public and private headers. The public header is not encrypted, whereas QUIC encrypts its private header along with the payload. The comparison of QUIC packet format with TCP + TLS is depicted as in Figure 1.5. QUIC public header contains connection ID and flags. The connection ID is an identifier used to identify a connection at an endpoint. The flags are used in the initial session establishment.

The remainder (private header and payload) is encrypted, so it is not visible to an eavesdropper. In TCP, the control of the protocol is explicit, so a third party (e.g. NOs, etc.) is able to inspect this information (e.g. sequence number, acknowledge number, etc.) for network troubleshooting. For example, NOs can infer network performance parameters (e.g. round-trip time, etc.) by inspecting TCP packets. In contrast, the private header and payload of QUIC are encrypted to protect from third-party inspection. This creates many obstacles for NOs in network troubleshooting. The detail is described thoroughly in section 1.3.

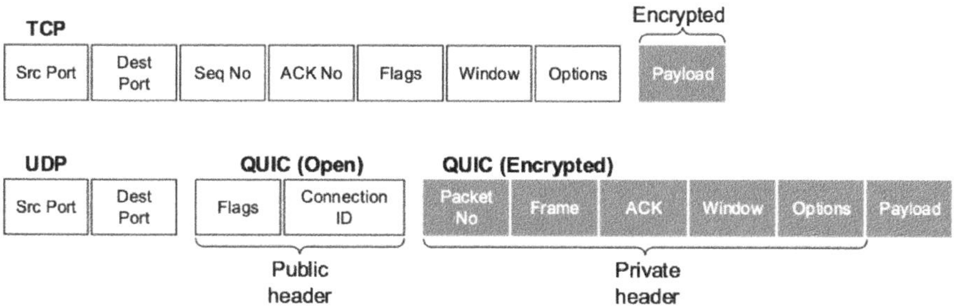

Figure 1.5. *Comparison of QUIC packet format with TCP + TLS. For a color version of this figure, see www.iste.co.uk/tong/troubleshooting.zip*

1.2.1.2. *Connection establishment*

Nowadays, many services require a secure and reliable connection. The combination of TCP and TLS is a potential solution for this purpose. TCP + TLS requires at least two RTTs to establish a secure connection, bringing a significant latency. QUIC integrates TLS1.3 to set up a secure connection with 0-RTT for a connection establishment time. In other words, encrypted payload is sent in the first packet when a previous connection is resumed. Comparison of connection establishment between QUIC and TCP + TLS is described as in Figure 1.6.

Concerning first-time connection establishment (Figure 1.6(a)), QUIC only uses 1-RTT by combining transport and cryptographic handshake in contrast to 3-RTTs in TCP + TLS1.2 and 2-RTTs in TCP + TLS1.3. QUIC combines both transport and cryptographic handshake in the first packet of connection to reduce latency in connection establishment. When a client connects to a server for the first time, it sends a Client Hello message to the server for key negotiation, along with other information (e.g. connection ID, preferred version number, etc.). The client encodes the handshake using the version number which is proposed. If the server does not support this version, it redirects the client to a version negotiation process. Otherwise, the server replies with a Server Hello message containing necessary information (e.g. certificate, session information, etc.) for subsequent connections. Next, the client can send encrypted packets to the server.

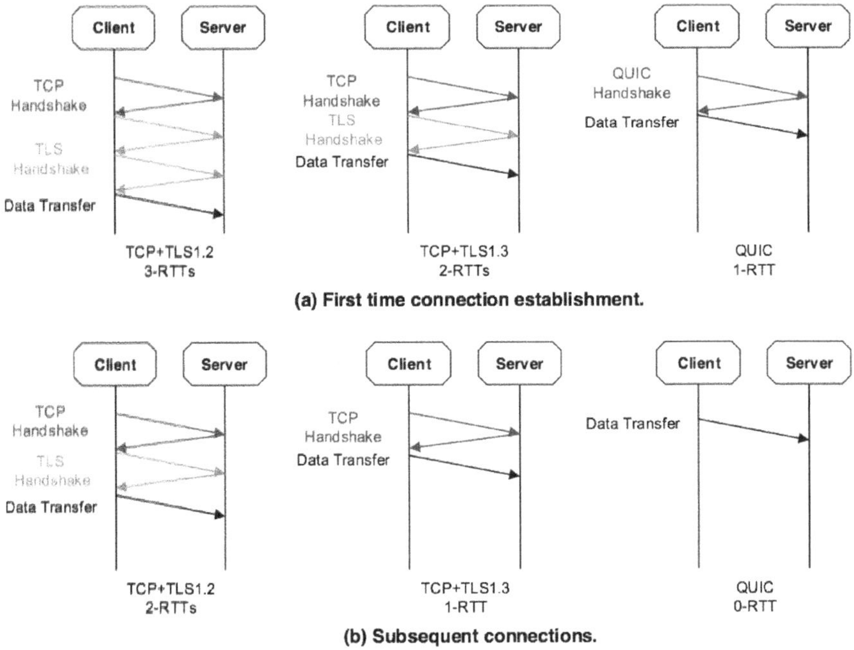

Figure 1.6. *Comparison of connection establishment between QUIC and TCP + TLS. For a color version of this figure, see www.iste.co.uk/tong/troubleshooting.zip*

The necessary information for the first-time connection establishment is stored in a cryptographic cookie at the client-side. It is used to authenticate the client in the subsequent connections. In the cookie, the server's Diffie–Hellman value is used to calculate an encryption key.

As for the subsequent connections (Figure 1.6(b)), the client sends the Diffie–Hellman value and its cookie to the server, along with the encrypted payload. The server uses information in the cookie to authenticate the client. Then, it uses the Diffie–Hellman value to calculate the encryption key to decrypt the encrypted payload and sends back the encrypted responses to the client.

1.2.1.3. *Multiplexing*

Multiplexing is a method of sending multiple data streams over a single connection. In HTTP1.1, a client can only request one resource in a single connection as in Figure 1.7(a), so it needs to open multiple concurrent TCP connections. However, each client has a limited number of connections to a server, so sending a new request over one of these connections has to wait until a previous connection is complete. This leads to the HTTP Head of Line blocking (HoL blocking), which brings additional latency and complexity of managing multiple connections.

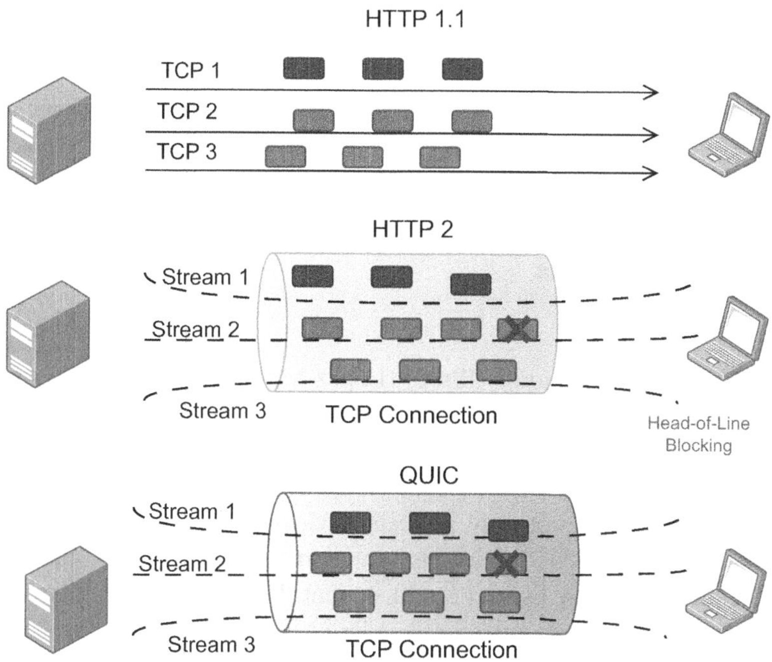

Figure 1.7. *Multiplexing comparison between HTTP1.1 and HTTP/2 over TCP and QUIC (Cui et al. 2017). For a color version of this figure, see www.iste.co.uk/tong/troubleshooting.zip*

HTTP/2 (Figure 1.7(b)) solves this problem by multiplexing multiple data streams over a single connection if there are multiple requests to the server. Data transmitted in the same connection is delivered to the client in order, leading to the TCP HoL blocking. If a TCP packet in a connection is lost, the server must wait until the TCP packet is retransmitted successfully before processing the following TCP packets.

QUIC (Figure 1.7(c)) supports multiplexing multiple data streams over a single connection without requiring ordered delivery of all packets.

Consequently, this can overcome the TCP HoL blocking. Data missing on a stream does not block the delivery of other streams.

1.2.1.4. *Connection migration*

TCP uses a port and an IP address of both endpoints to identify a connection. All network connections will be interrupted when a client changes its IP address such as moving out of a Wi-Fi network range or switching from a wired to a cellular network. However, QUIC uses a connection ID (identifier) randomly generated by a QUIC client to identify a connection. When a QUIC client changes its IP address, it can continue to use the current connection ID from a new IP address without any connection interruptions.

1.2.2. *Other protocols*

1.2.2.1. *IPsec*

Internet Protocol Security (IPsec) (Lopez-Millan et al. 2019) is a secure network protocol that encrypts and authenticates data packets to provide secure communication between two endpoints in the network. IPsec is used widely in virtual private networks (VPN). IPsec is a layer 3 end-to-end security scheme, so it not only protects payload but also IP header of data packets. IPsec uses UDP packets on port 500 for initial handshake, authentication and shared secret establishment. In this protocol, there are two main parts including authentication header (AH) and Encapsulating Security Payload (ESP). In the infancy of IPsec, the former is responsible for authentication while the latter provides data confidentiality. ESP adds a header and a trailer to each packet as in Figure 1.8. IPsec can operate in two modes: transport and tunnel modes. Only data and ESP trailer are encrypted in the transport mode, whereas an entire original packet and ESP trailer are encrypted in the tunnel mode.

1.2.2.2. *TLS*

TLS Rescorla and Dierks (2018) is a successor of Secure Sockets Layer (SSL), a cryptographic protocol designed to provide a communication security in the network. It runs in the application layer to provide privacy and data integrity in data communication. TLS is considered as a secure layer in Hypertext Transfer Protocol Secure (HTTPS), a popular secure communication protocol on the Internet.

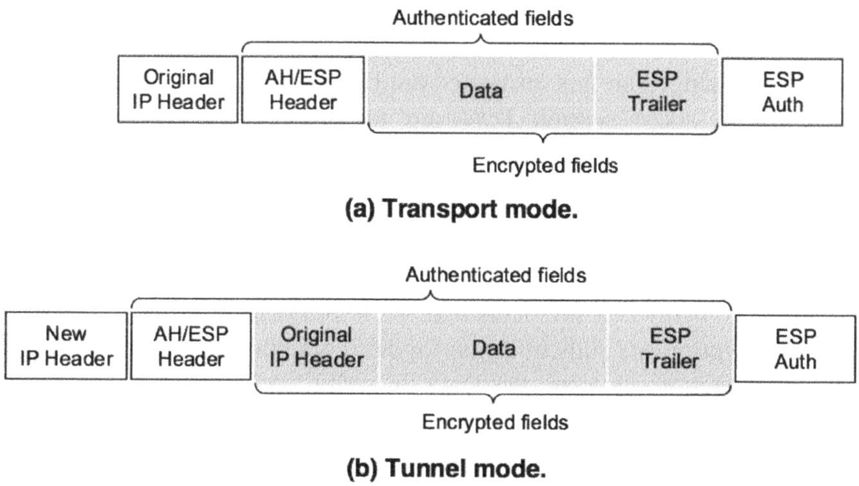

Figure 1.8. *IPsec packet structure. For a color version of this figure, see www.iste.co.uk/tong/troubleshooting.zip*

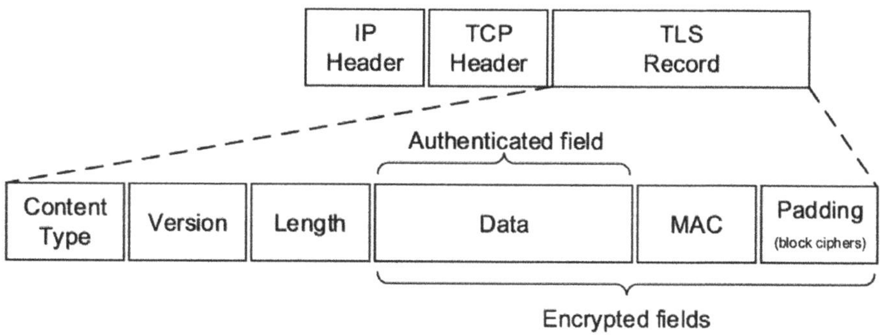

Figure 1.9. *TLS record packet*

TLS contains two main layers including TLS record and TLS handshake. The former one aims to divide data into compressed fragments. In a TLS record, a fragment is followed by message authentication code (MAC). The fragment and its MAC are encrypted together in a TLS record. The structure of TLS record packet is described as in Figure 1.9. During the TLS handshake, clients and servers exchange messages to acknowledge and verify each other. Then, it establishes an encryption algorithm and agrees on session keys. At the beginning of TLS connections, clients and servers are authenticated according to X.509 certificates chain. In this phase, TLS messages exchanged are not encrypted until session keys are established and agreed. Next, these session keys are utilized by the TLS record to communicate data.

1.3. Drawbacks of troubleshooting with encrypted traffic

Nowadays, many new protocols (e.g. TLS1.3, QUIC, etc.) are designed to provide encryption and authentication for data communication in the network. The authentication protects the data integrity and prevents unexpected modifications in data packets. The encryption provides the confidentiality of transport data and prevents a data inspection for illegal purposes (e.g. stealing information, etc.). This reduces the ability of NOs to obtain an insight into their networks and effectively manage their networks. In this section, we present two drawbacks of network troubleshooting in the context of encrypted traffic.

1.3.1. *Network performance monitoring*

Network performance monitoring is an essential part of the data collection module to collect network parameters for network troubleshooting. Parameter measurement helps NOs to troubleshoot problems and identify whether the problems happen in their network or other sides (e.g. service providers, clients, etc.). Measurement approaches monitor network traffic and use information in packets (e.g. sequence number, acknowledgment number, etc.) to measure network parameters (e.g. quality of experience (QoE), etc.) or additional information (e.g. application class, etc.).

1.3.1.1. *QoE estimation*

Concerning the QoE, many existing studies calculate QoE using a QoE model based on subjective data (Amour et al. 2020). The QoE model is built according to application-level parameters (e.g. start-up delay, number of stalling events, bitrate, resolutions, etc.). These parameters can be extracted by analyzing client or server-side logs, but NOs do not have access to this information. Hence, NOs can implement network-side measurement approaches by analyzing network traffic to obtain these parameters. In fact, the application-level parameters can be measured thanks to the information in the HTTP packet headers (Schatz et al. 2012; Casas et al. 2014). For example, YouTube's QoE is determined by stalling events in the video playback. Therefore, Schatz et al. (2012) proposed a deep packet inspection technique to measure the number of stalling events using IP network flows. The objective is to measure the playing time accumulated in the buffer of the YouTube player by comparing timestamp of packets and playback time of video frames. When the playback buffer is empty, the video starts to stall. The playback time can be obtained according to the HTTP packet headers. However, service providers (e.g. YouTube, etc.) nowadays encrypt network traffic (e.g. using TLS, QUIC, etc.) to protect user's privacy and data during data transmission. Consequently, QoE estimation approaches using the inspection of HTTP packet headers are no longer available. Therefore, Orsolic et al. (2017) and Orsolic and Skorin-Kapov (2020) proposed a QoE classification approach for YouTube traffic using a machine learning (ML) algorithm. This approach monitors and analyzes network traffic to obtain 54 network features divided into six categories: packet length, transferred data size, packet count, inter-arrival time, throughput and TCP flag count. Then, these parameters are analyzed in the ML algorithm to classify QoE into three QoE classes: low, medium and high QoE. There are many ML algorithms, so this approach evaluates performance of several ML algorithms (OneR, Naive Bayes, SMO, J48 and Random Forest) to select an appropriate one for the QoE classification.

1.3.1.2. *Application identification*

Nowadays, with the development of 5G, there are many strict SLA requirements from end-users (e.g. high reliability, low latency, etc.) (Lin et al. 2019). Consequently, differential treatment is expected by end-users. The emerging of application-aware mechanisms (e.g. network slicing, traffic engineering, etc.) which implement different policies corresponding to various applications, is a potential solution to overcome these drawbacks. In the past, NOs used port numbers or signatures in payload of packets to identify the application classes Lopez-Martin et al. (2017). Libprotoident (Finsterbusch et al. 2013) is an open-source software library that implements the deep packet inspection for protocol identification. Libprotoident only uses port numbers and the first four bytes of payload in each direction. For example, Libprotoident identifies HTTP and SIP flows according to common strings (e.g. ET, PUT, HTTP, SIP, INVI, REGI, etc.) and registered ports

(80, 8080, 8081, 443). Similarly, Adami et al. (2015) implement a deep packet inspection approach to identify the VoIP flows using signatures of SIP protocol (e.g. content-type: "application/sdp", etc.). However, the packet's payload can be encrypted with encryption protocols (e.g. TLS, etc.). Therefore, Deri et al. (2014) proposed nDPI, an open-source library for protocol identification using the information in both header and payload of packets. nDPI can handle TLS-encrypted traffic to identify known applications. In this case, only the network traffic in the initial key exchange can be decoded. nDPI can extract the hostname of servers to identify different applications. For instance, network flows toward the server name "api.facebook.com" are assigned as Facebook application. However, the server name can be encrypted according to DNS over TLS, HTTPS or QUIC Huitema and Rescorla (2018). Consequently, encrypted traffic classification approaches using packet inspection are not effective. This results in many obstacles for the encrypted traffic classification. Amaral et al. (2016) proposed a new traffic classification method using flow-based features and an ML algorithm. This method monitors and analyzes network traffic to extract 12 flow-based features from the first five packets of flows (e.g. packet size, packet time-stamp, inter-arrival time, etc.). Then, these features are analyzed in the ML algorithm to identify the application classes. Lopez-Martin et al. (2017) proposed a traffic classification approach using flow-based features and a deep learning (DL) algorithm to identify 108 application classes for Internet of Things (IoT) network.

1.3.2. *Intrusion detection system*

1.3.2.1. *Firewall*

In the network, middlebox is deployed to block malicious traffic and prevent illegal activities. It can classify network traffic into malicious or benign traffic to make appropriate policies (Kühlewind et al. 2018). In the TLS-encrypted traffic, Service Name Indicator (SNI) is used to identify malicious connections. If SNI in the Client Hello TLS message matches Subject Alternate Name (SAN) in the server certificate, the connection is considered as a benign connection. However, SNI can be encrypted to protect the information about destinations (e.g. domain name of websites, etc.) (Huitema and Rescorla 2018). Nowadays, SNI is encrypted in the Domain Name System (DNS) over TLS, HTTPS and QUIC.

Besides, the middlebox can be used for content filtering (Moriarty and Morton 2018). The middlebox is used to block content to meet requirements from law enforcement or regulatory authorities (e.g. enforcing content-based billing, etc.). They analyze data packets to obtain URL (Uniform Resource Locator) field. If it matches a blocked URL in a blacklist, a 404 error notification is returned. However, these approaches are unavailable when traffic is encrypted. DNS over TLS, HTTPS and QUIC hide the DNS information in the packets.

When network traffic is encrypted, it results in obstacles for malicious traffic prevention and content filtering. If the middlebox cannot inspect network traffic for network management, it needs to be moved to the endpoints at a higher operational cost.

1.3.2.2. *DDoS protection and migration*

Distributed denial-of-service (DDoS) protection is necessary in network operations. It aims to detect DDoS traffic and redirect it to a migration system to filter out DDoS traffic from legitimate traffic.

DDoS can be implemented according to a botnet, a collection of compromised machines or bots (Antonakakis et al. 2012). Many bots nowadays use a domain generation algorithm (DGA) to generate domain names and connect to a Command and Control (C&C) server. Therefore, many studies analyze the domain name of C&C server to detect DGA botnet and prevent DDoS attacks. Woodbridge et al. (2016) proposed a DGA botnet detection system using a long short-term memory (LSTM) network. This system monitors and analyzes DNS packets to obtain the domain name of C&C server. Then, the domain name is analyzed in the LSTM network to detect the DGA botnet. If a domain name is classified as a malicious domain, the IP address of C&C server corresponding to this domain is blocked to prevent DDoS attacks. Tran et al. (2018) proposed LSTM.MI, a LSTM-based DGA botnet detection framework for handling multiclass imbalance. In LSTM.MI, original LSTM is adapted to be cost sensitive to handle the multiclass imbalance. However, DNS information (e.g. domain name, IP address, etc.) is hidden with DNS over TLS, HTTPS and QUIC. This leads to many challenges for the DGA botnet detection systems as well as DDoS prevention.

Besides, DDoS can be detected according to a signature-based intrusion detection system (IDS) (Shurman et al. 2019; Otoum and Nayak 2021). This approach inspects network traffic and checks anomalous signatures in a predefined list of signatures such as file hashes, malicious domains, known byte sequences, etc. For example, Snort AbuHmed et al. (2008) is an open-source IDS for inspecting intruder signatures. Snort rules contain protocol, source IP address, destination IP address, source port, destination port and signatures in the payload for one or more patterns. When network traffic is encrypted, this approach is ineffective. In this case, network features (e.g. packet length, inter-arrival time, etc.) can be used to detect DDoS traffic. Garcia et al. (2021) proposed an anomaly detection system using Artificial Intelligence (AI) for detecting DDoS traffic over encrypted traffic. This approach monitors network traffic and extracts 57 network features (e.g. packet size, number of bytes, etc.). Then, these features are analyzed in a detection model to classify network traffic into legitimate or intrusive. In the detection model, network features are first analyzed in a Gaussian mixture model to calculate a probability density. If it is less than *threshold1*, network traffic is assigned as intrusive. If it is bigger than *threshold2*, it is assigned as legitimate. Otherwise, a reconstruction error is calculated according to Auto-Encoder. If the reconstruction error is less than *threshold3*, it is assigned as legitimate. Otherwise, it is assigned as intrusive. Similarly, Zolotukhin and Hämäläinen (2018) proposed an application-layer DDoS detection method for encrypted traffic. In this method, there are two main phases containing training and detection phases. In the training phase, this method monitors network traffic to collect network features of seven categories: duration of conversation, number of packet sent in 1 s, number of bytes sent in 1 s, average packet size, average TCP window size, average TTL and average of packets with different TCP flags. Then, legitimate traffic is clustered into clusters using an ML algorithm (e.g. K-mean, K-medoids, Fuzzy c-means, etc.). In the detection phase, network traffic is assigned as intrusive when it does not belong to any clusters.

1.4. Conclusion

With the ever-increasing of network devices, computer networks have become more and more complex. Therefore, network troubleshooting plays an important role in network systems (Fonseca and Mota 2017; Yu et al. 2018; Cherrared et al. 2019). This section surveyed related work on several network problems: rule failure, link failure and switch's buffer overload. Facing a growing number of data theft attacks, many service providers encrypted network traffic to protect data and user's privacy. The network traffic can be encrypted by encryption protocols such as IPsec, TLS, etc. This leads to many obstacles for network troubleshooting for encrypted traffic related to network monitoring approaches and intrusion detection systems.

When the packet's payload is encrypted (e.g. TLS, etc.) or even entire packet is encrypted (e.g. IPsec in the tunnel mode, etc.), it prevents NOs from inspecting network traffic to measure network parameters (e.g. QoE, etc.). Remarkably, the information about the application class is hidden in this case. Consequently, many encrypted traffic classification techniques are studied to identify the applications without decrypting the network traffic. Besides, encrypted traffic results in many difficulties for intrusion detection systems. In the past, IDS detects different kinds of attacks according to its signatures in the network traffic. However, these signatures are hidden due to encrypted traffic. Consequently, IDS today detects different kinds of attacks because of network parameters (e.g. packet length, inter-arrival time, etc.).

The traditional troubleshooting architecture has been studied by the research community over the past decades. Nevertheless, in the context of encrypted traffic, this architecture shows limitations related to network performance monitoring approaches and intrusion detection systems. In the next chapter, we present a novel troubleshooting architecture for encrypted traffic and the data collection module, which aims to collect data for further modules in network troubleshooting.

2

Novel Global Troubleshooting Framework fo Encrypted Traffic

"The extension of the encryption boundary into the transport layer"

Mirja Kühlewind, MAMI Management and Measurement Summit 2018

In Chapter 2, we present a novel troubleshooting architecture in the context of encrypted traffic and then present a proof of concept of this architecture in a software-defined networking (SDN) environment. We also present the data collection module that helps to build datasets for further modules in network troubleshooting.

2.1. Novel network troubleshooting architecture for encrypted traffic

As discussed in Chapter 1, the traditional troubleshooting architecture designed for non-encrypted traffic contains four essential modules: *data collection, anomaly detection, root cause analysis* and *remediation*. In this architecture, after detecting anomalies in the anomaly detection, its root cause will be identified in the root cause analysis (RCA) module. Finally, the root causes will be addressed in the remediation module to return to a normal state in the network. During the implementation of the RCA and remediation modules, the network will be negatively influenced (e.g. high latency, low reliability, etc.), resulting in negative economic impacts for NOs. According to a report on the North American Network Operator Group (Zeng et al. 2012b), the duration time for the RCA and remediation approaches can be from 1 h to more than 5 h depending on the status of anomalies. In fact, YouTube lost $34,000 due to 0.17 h of downtime, while Cisco lost $1,066,000 because of 5.33 h of downtime in their cloud servers (Gagnaire et al. 2012).

Nowadays, many service providers (e.g. YouTube, Facebook, etc.) encrypt their data during transmission to protect the data and the user's privacy. However, the traditional troubleshooting architecture was not designed for encrypted traffic. According to related work (Kühlewind et al. 2018; Moriarty and Morton 2018), encrypted traffic results in many obstacles related to network performance monitoring approaches in data collection (e.g. QoE estimation, application identification, etc.) and intrusion detection systems for anomaly detection. When traffic is encrypted, the information in the packet header and payload are invisible. Hence, from the point of view of network operators, network performance monitoring approaches that implement a deep packet inspection face many obstacles to calculate the QoE or identify the application classes (Schatz et al. 2012; Deri et al. 2014; Lopez-Martin et al. 2017). Moreover, intrusion detection systems that depend on signatures in the packet's payload (e.g. file hashes, known byte sequences, etc.) to identify different kinds of attacks are no longer available (Shurman et al. 2019; Otoum and Nayak 2021).

Therefore, in this book, we present a novel troubleshooting architecture for NOs in the context of encrypted traffic. In this architecture, we present two processes after detecting an anomaly in the network. The first process is the temporary remediation module that ensures the availability and reduces the affection of anomalies in the network until the root causes are addressed. In parallel with this process, we also implement the second process that is responsible for identifying and dealing with the root causes. The aim when implementing two processes is to reduce the negative impacts of network problems during the RCA and definitive remediation.

Concerning the drawbacks of network troubleshooting, using network parameters is a potential solution for network performance monitoring approaches (e.g. QoE estimation, application identification, etc.) and intrusion detection systems in the context of encrypted traffic. In fact, there are DDoS detection methods for encrypted traffic using network parameters (e.g. packet length, inter-arrival time, etc.) (Garcia et al. 2021; Zolotukhin and Hämäläinen 2018). Similarly, Orsolic et al. (2017) and Orsolic and Skorin-Kapov (2020) proposed a QoE classification technique for YouTube traffic using 54 network parameters and ML algorithms. To identify the application classes, encrypted traffic classification approaches are using ML algorithms. These approaches focus on HTTPS (Wang et al. 2018) and VPN traffic (Wang et al. 2017). Consequently, we present an encrypted traffic classification approach to identify application classes for QUIC traffic. Application identification plays an important role in application-aware remediation mechanisms in network troubleshooting.

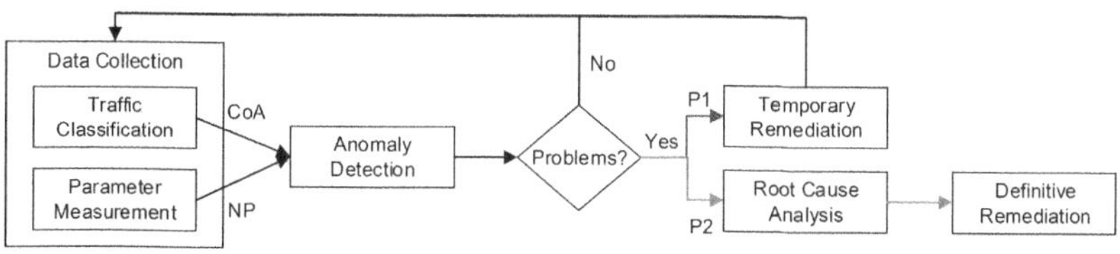

Figure 2.1. *The novel troubleshooting architecture in the context of encrypted traffic. For a color version of this figure, see www.iste.co.uk/tong/troubleshooting.zip*

Figure 2.1 shows the novel troubleshooting architecture designed for encrypted traffic. The major modules are described as follows:

– *Data collection*: this is composed of two essential modules: traffic classification and parameter measurement. The former aims to identify the class of application (CoA) on network flows in the context of encrypted traffic, while the latter aims to monitor and collect network parameters (NP) for further modules.

– *Anomaly detection*: this module takes into account time-series network parameters from data collection to detect an anomaly (abnormal symptom of network problems) in the network. The purpose is to identify the moment when problems happen in the network by detecting its symptoms.

– *Temporary remediation*: this module implements remediation approaches to temporarily reduce the negative impacts of an anomaly (e.g. high latency, high loss, etc.) until its root causes are identified and solved.

– *Root cause analysis*: an anomaly can be caused by many root causes. For example, congestion can be caused by link failure, switch failure, buffer overload, etc. This module aims to identify its root causes (the problem leading to the anomaly) for further modules.

– *Definitive remediation*: after identifying the root causes of an anomaly, this module is used to solve its root causes completely.

2.2. Proof of concept of novel troubleshooting architecture in SDN

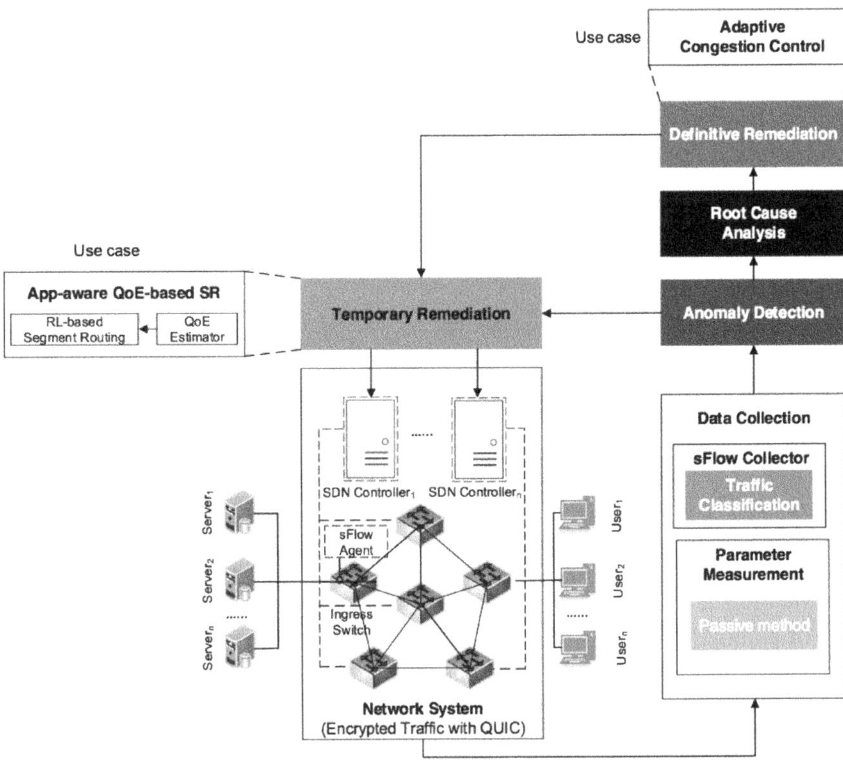

Figure 2.2. *The novel troubleshooting framework in the SDN environment. For a color version of this figure, see www.iste.co.uk/tong/troubleshooting.zip*

To solve root causes (e.g. switch failure, misconfiguration, etc.) in the network, we sometimes need to update network devices. In traditional network architecture where control logic is distributed, the policies in the network troubleshooting are updated separately in each network device (Bannour et al. 2017). This can lead to inconsistency between network devices that results in other problems in the network. Fortunately, SDN has emerged as a potential solution to overcome this obstacle. SDN with a separation between the control and infrastructure layers offers global visibility and better flexibility to troubleshoot the network. SDN also offers monitoring tools (e.g. Link Layer Discovery Protocol) to facilitate network troubleshooting. Consequently, the troubleshooting architecture is implemented in the SDN environment. In this context, we consider ONOS (Berde et al. 2014) as an SDN controller due to its popularity in existing studies (Bannour et al. 2017, 2020).

As a proof of concept of the troubleshooting architecture for encrypted traffic, we present a troubleshooting framework for encrypted traffic in the SDN environment. This framework is described as shown in Figure 2.2 with five main components: *data collection*, *anomaly detection*, *temporary remediation*, *root cause analysis* and *definitive remediation*. During the transmission from servers to clients, network flow is transmitted from ingress switches to a sFlow collector (Ujjan et al. 2020) to obtain the class of application because of the traffic classification module. Also, the measurement module monitors this flow to collect network parameters. After that, these parameters are analyzed in the anomaly detection module to detect anomalies (e.g. increase of latency, packet loss, etc.). When anomalies occur in the network, we need to implement the temporary remediation module to guarantee the availability of the network and meet strict SLA requirements. There are many remediation approaches, including load balancing, routing, etc. Many studies proved the performance of routing in problem remediation (Kim et al. 2016; Jurkiewicz et al. 2020). In fact, SR (segment routing) is implemented by network operators (e.g. NTT (Cisco 2020), Vodafone (Cisco 2018), etc.). Therefore, we present an application-aware QoE-based SR in the temporary remediation module. In parallel with the temporary remediation module, we also need to implement the RCA module to identify the root cause of anomalies and solve it completely in the definitive remediation module. There are many problems in the network (e.g. misconfiguration, security, congestion, etc.), so we present a use-case related to the RCA and definitive remediation approach in this book. Here, we consider congestion to generate anomalies because it is a widespread issue from the point of view of NOs (Zeng et al. 2012b). Concretely, the development of IoT and 5G increases the amount of Internet traffic, introducing much more pressure on the network infrastructure, resulting in congestion in the network, particularly during the confinement of Covid-19. Congestion can be caused by link failure, switch failure or buffer overload. The objective of the RCA is to identify the root causes of congestion. Concretely, the RCA is used to identify the type of problems (link failure, switch failure or buffer overload) leading to congestion in the network. If the link or switch failure happens in the network, we will send notifications to administrators to address it. Otherwise, we present an adaptive congestion control mechanism in the definitive remediation to deal with buffer overload. The major components in this framework are described as follows:

– *Data collection* consists of two essential modules: traffic classification and parameter measurement. In the parameter measurement module, we measure network parameters such as latency, packet loss, link utilization, number of packet sent, number of packet received, number of byte sent, number of byte received, number of flow entries in switch and QoE. These parameters are measured according to LLDP (Liao and Leung 2016; Li et al. 2018) and *PortStatistics* (API Team 2021) in SDN controllers. The purpose is to build troubleshooting datasets in static and dynamic network for further modules. In the traffic classification module, there are two main stages of classification. The first classification stage is to identify chat and VoIP applications using flow-based features and an ML algorithm. The second classification stage aims to identify video streaming, file transfer and Google play music applications using packet-based features and a convolutional neural network. To implement this module, we need to collect raw network traffic. Collecting network traffic from OpenFlow switches is not effective with a huge amount of network traffic. Therefore, we implement sFlow (Ujjan et al. 2020), an industry standard supporting a packet sampling technique, to collect 10% of network traffic at ingress switches. Concretely, sFlow agents sample network traffic traversing the ingress switches and forward it to the sFlow collector for the traffic classification.

– *Anomaly detection* takes into account time-series network parameters to detect an abnormal symptom of network problems. This module aims to identify the moment when problems occur in the network by detecting its symptoms (anomalies). To generate network anomalies, we use a fault injection technique (Cid-Fuentes et al. 2018; Hong et al. 2020) to generate congestion to the network. We implement an anomaly detection approach using three time-series network parameters (latency, packet loss and link utilization) and an ML algorithm. To choose an appropriate ML algorithm, we evaluate the performance and processing time of ML algorithms such as support vector machine, random forest, Adaboost and convolutional neural network.

– *Temporary remediation* is the application-aware QoE-based segment routing mechanism that aims to consider various segment routing strategies corresponding to different kinds of applications to meet strict SLA requirements. In a specific routing strategy, this module monitors network states and selects appropriate paths using reinforcement learning. The segment routing mechanism is based on the feedback of the network environment (QoE) to adapt to unexpected network conditions. The segment routing mechanism is evaluated according to an emulation testbed with the SDN controller *ONOS* and the emulator *mininet*. We also customize the *org.onosproject.segmentrouting* application (Berde et al. 2021) supported by *ONOS* to implement the segment routing mechanism. In reinforcement learning, a trade-off between exploration and exploitation phases needs to be considered thoroughly. Therefore, it is formalized as a multi-armed bandit problem. In this book, we evaluate the performance of three selection algorithms (*ϵ-greedy, softmax* and *UCB1*) in four scenarios (perfect scenario, delay scenario, loss scenario and scenario with both delay and loss) to select an appropriate one for solving the multi-armed bandit problem. The segment routing mechanism is evaluated with benchmarks (*Standard_SR* and *Max_QoE* mechanisms) related to mean opinion score (MOS) against time, average optimal MOS, median MOS, 95% confidence interval of MOS, CPU usage and control overhead.

– *Root cause analysis* aims to identify the root cause of congestion in the network. The objective is to identify the type of problems resulting in congestion (link failure, switch failure or buffer overload). We implement an ML-based RCA using nine time-series network features (e.g. latency, packet loss, number of byte sent, number of packets sent, etc.). The performance of ML algorithms (support vector machine, bagging, random forest, Adaboost, gradient boosting and convolutional neural network) is evaluated to select an appropriate one. We implement a feature selection method (wrapper method) to identify an appropriate feature set to obtain a good performance. The performance of RCA is evaluated with datasets in static and dynamic networks.

– *Definitive remediation* is an adaptive congestion control mechanism that is in charge of dealing with root causes of congestion completely to return to a normal state in the network. In this module, we present an adaptive BBR (bottleneck bandwidth and round-trip time) algorithm using reinforcement learning for QUIC traffic. This algorithm aims to adjust the sending rate at sender sides corresponding to different network conditions. Adaptive BBR and benchmarks are evaluated in the context of HTTP/3 in contrast to existing studies evaluating congestion control algorithms in the context of HTTP/2. We customize the open-source *lsquic* (Dmitri and George 2021) to implement the adaptive BBR algorithm and benchmarks. The performance of these algorithms is evaluated in an emulation testbed in a dynamic network where the RTT and loss of link change after a given time.

2.3. Data collection

As discussed in the previous section, data collection plays a vital role in network troubleshooting. This is explained by the fact that the anomaly and its root causes result in a fluctuation of data collected in the network. In this section, we present several kinds of data classification, monitoring tools, and the parameter measurement module, which collects data for network troubleshooting.

2.3.1. *Data classification*

According to related work (Eberendu 2016; Amour et al. 2015), data can be classified in different methods as follows:

– data types (Eberendu 2016): structured, unstructured and semi-structured data;

– source (Amour et al. 2015): application-level data, network-level data, user feedback, user profile, etc.;

– quantity: large data, which need to be processed by big data processing techniques (e.g. Hadoop, etc.) before being taken into account, and small data, which can be taken into account without big data processing techniques.

In this chapter, we consider the data classification according to data types widely used in network troubleshooting (Cherrared et al. 2019; Du et al. 2017; Revina et al. 2020).

Structured data refer to data that contain a definite format (e.g. network parameters, etc.). Therefore, it is easy to query and retrieve necessary information.

On the contrary, unstructured data contain no particular structure (e.g. troubleshooting ticket, etc.). The troubleshooting ticket is used to identify the anomaly and diagnose its root causes. When problems occur in the network, clients contact an IT helpdesk to report their problems. Then, the IT helpdesk collects information from the client's reports, generates a troubleshooting ticket and assigns it to the corresponding processing team to diagnose. However, troubleshooting tickets contain a large amount of unstructured text data, leading to many obstacles for data parsing and processing (Revina et al. 2020).

The semi-structured data are a combination of structured and unstructured data (e.g. system logs, etc.). System logs record various system states and alarms to identify performance issues, failures and root causes. The information in the system log contains structure data including the IP address of the object generating an alarm, a timestamp, an alarm identifier, the severity level of failures, etc. (Du et al. 2017; Cherrared et al. 2019). The system log also contains unstructured data due to the existence of network issues. The system log is available in the network systems, so it is a valuable resource for understanding network status. In fact, there are many anomaly detection approaches using system logs for identifying the anomaly in the network (Yu et al. 2016; Du et al. 2017). However, these approaches contain several limitations including (Du et al. 2017; Cherrared et al. 2019):

– Unstructured data: log data are unstructured, and it varies significantly between different network systems. Therefore, it is challenging to parse log data in the distributed network systems with different kinds of network devices.

– Processing massive data: log data are used to identify the anomaly in the network, so it needs to be processed and analyzed in real time. However, log data is usually recorded in a huge file (e.g. a few GB, etc.). Therefore, processing massive data in a short time is a challenge.

– Desynchronization: the clock desynchronization of systems logs in distributed systems can influence the order of received alarms and result in uncertain results.

In this book, we consider the network performance parameters as data for network troubleshooting. In fact, it is studied by the research community in network troubleshooting (Fonseca and Mota 2017; Yu et al. 2018; Cherrared et al. 2019). In the following subsection, we will present network monitoring techniques for data collection in network troubleshooting.

2.3.2. *Monitoring tools*

To collect network parameters, there are two monitoring approaches including active and passive monitoring. Active monitoring approaches (Sinha et al. 2015; Atary and Bremler-Barr 2016) send probe packets into the network systems and analyze feedback to obtain network parameters. Passive approaches monitor the network traffic and analyze it to obtain the network parameters. The active approaches generate overhead in the network, making the status of problems more grave. Moreover, a loss of probe packets can lead to uncertain results. Therefore, we consider passive monitoring approaches to collect network parameters. In fact, there are popular standards for network monitoring including Simple Network Management Protocol (SNMP), NetFlow, sFlow and OpenFlow-based monitoring approaches as follows:

– SNMP is a popular protocol in network management, collecting information from different network devices (e.g. routers, switches, etc.) in the network (Al-Naymat et al. 2018). SNMP offers massive statistical information about network devices through a database, called the management information base (MIB). MIB is a collection of parameters reflecting physical links (e.g. maximum packet size, transmission rate, etc.), routing (e.g. routing metrics, destination address, etc.), underlying protocols (e.g. TCP, ICMP, etc.) and so on. However, SNMP contains several drawbacks (Amirthalingam and Moorhead 1995). The first one is unreliability. SNMP uses the underlying User Datagram Protocol (UDP), so there is no guarantee of delivery, ordering or duplicate protection for data transmission. The second one is related to a simple data structure. SNMP only supports a simple data structure, so it is inefficient to transfer data in a short time.

– NetFlow (Suárez-Varela and Barlet-Ros 2017) is a flow sampling technique designed by Cisco, which is supported by many vendors including Juniper, HP, OpenVswitch, etc. NetFlow samples a flow with a specific probability, aggregates all packets of this flow into flow records and then exports toward NetFlow collectors. The NetFlow architecture is depicted in Figure 2.3. NetFlow contains three main components including the NetFlow exporter, NetFlow collector and analysis application. NetFlow exporter aggregates all packets of a flow into flow records (with a probability) and exports to one or more NetFlow collectors. NetFlow collector pre-processes data from NetFlow exporter and stores it in NetFlow storage. Analysis application collects data from NetFlow storage and analyzes for the network management.

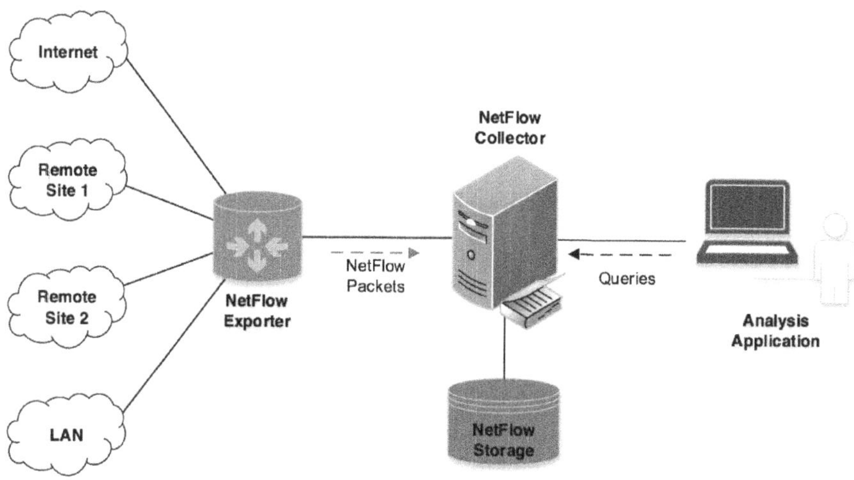

Figure 2.3. *NetFlow architecture (Suérez-Varela and Barlet-Ros 2017). For a color version of this figure, see www.iste.co.uk/tong/troubleshooting.zip*

– sFlow (Ujjan et al. 2020) is a standard for network monitoring with a packet sampling technique that is supported by vendors containing IBM, HP, OpenVswitch, etc. sFlow samples each packet with a specific probability, aggregates sampled packets into sFlow records and exports to sFlow collectors. The sFlow architecture is shown in Figure 2.4.

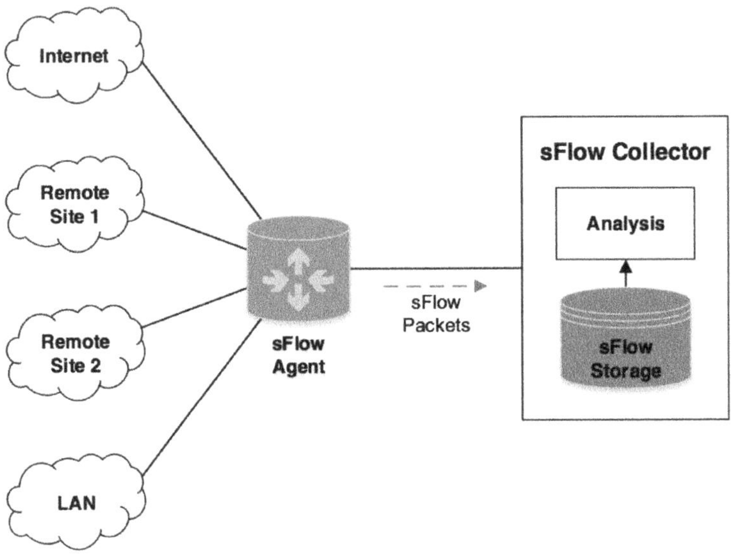

Figure 2.4. *sFlow architecture. For a color version of this figure, see www.iste.co.uk/tong/troubleshooting.zip*

– OpenFlow-based monitoring approach has been studied by the research community. Similar to NetFlow and sFlow, OpenFlow-based monitoring approaches collect network parameters based on information provided by OpenFlow agents in OpenFlow-supported devices. OpenFlow (Lara et al. 2013) is a communication protocol providing access to the forwarding plane of network devices (e.g. switch, router, etc.) in the network. Consequently, it offers opportunities to overcome the limitations of monitoring standards (e.g. sFlow, NetFlow, etc.).

In comparison with OpenFlow-based approaches, sFlow and NetFlow have several limitations related to scalability, flexibility and reliability issues. Regarding the scalability issue, these approaches require configuring an agent for each network device. As for the flexibility issue, the agents send data from all network devices to the collectors periodically, which does not allow the selection of specific devices to be observed. Finally, concerning the reliability issues, these approaches use the UDP protocol (a connectionless protocol) for data transmission, so lost UDP packets can lead to uncertain results. The difference between NetFlow, sFlow and OpenFlow-based monitoring approaches is described in Table 2.1.

	NetFlow/sFlow	OpenFlow-based approaches
Scalability	Requires the configuration of an agent for each network device	Agent is configured automatically according to OpenFlow protocol
Flexibility	Send data from all devices to collectors periodically	Select which device is observed at any time slot
Reliability	Unreliability due to using underlying UDP to send data to collectors	Use Southbound API

Table 2.1. *Comparison between NetFlow, sFlow and OpenFlow-based monitoring approaches*

2.3.3. *Parameter measurement*

In this book, we consider a dataset consisting of context-dependent parameters including:

- latency;
- packet loss;
- link utilization;
- number of packets sent;
- number of packets received;
- number of bytes sent;
- number of bytes received;
- number of flow entries in switch;
- quality of experience (QoE).

2.3.3.1. *Latency*

Latency is calculated according to the Link Layer Discovery Protocol (LLDP) in SDN controllers, as in existing studies (Liao and Leung 2016; Li et al. 2018). In SDN controllers, LLDP is used to discover and update network topology. The detail is depicted in Figure 2.5. To update the status of the link s_1–s_2, the controller sends a Packet-out message containing an LLDP packet to instruct switch s_1 sending this LLDP packet to the neighboring switch s_2. The switch s_2 receives the LLDP packet, encapsulates this packet in a Packet-in message and sends it back to the controller. To estimate the latency, the controller encodes the *time-stamp* and puts it in the LLDP packet of the Packet-out message. When the controller receives the Packet-in message, it will decode it to obtain the *time-stamp*. The latency is the difference between the current time and the *time-stamp*.

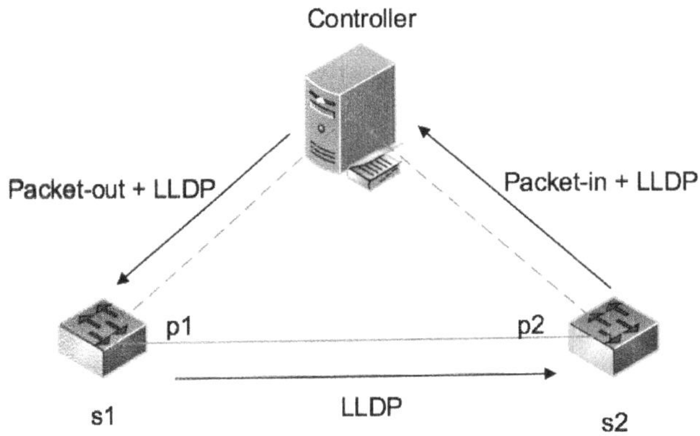

Figure 2.5. *The link discovery in LLDP. For a color version of this figure, see www.iste.co.uk/tong/troubleshooting.zip*

2.3.3.2. *Packet loss*

Packet loss refers to the loss of LLDP packets on each link. When the controller sends the Packet-out message, we store the *time-stamp* of each LLDP packet in this message in a database. After the Packet-in message arrives at the controller, we check the *time-stamp* and remove this from the database. After a given time, we increase the number of packet lost on a particular link by one if the difference between the current time and *time-stamp* in the database is bigger than a threshold. Then, we remove the *time-stamp* from the database. The above parameters are calculated on each link and then aggregated to calculate parameters on the routing path of network flows.

2.3.3.3. *Quality of experience*

QoE is a measure of user satisfaction of multimedia services (e.g. video streaming, file transfers, etc.). The mean opinion score (MOS) is used to estimate the QoE on the user's side. Therefore, it is necessary to have a subjective dataset related to MOS. Consequently, we use an MOS dataset built in a collaboration between our laboratory (LISSI) and Orange (a Network Operator in France). This dataset was built in a subjective test campaign to obtain 1,560 samples covering 23 parameters (e.g. application parameters, network parameters, etc.). The testbed for this campaign was implemented in the LISSI laboratory where 181 participants were asked to watch a short video and give their opinions. These participants are researchers and students from different disciplines between 19 and 38 years, with little or no experience in video assessment experimentation. There are five levels of participants' opinions including 1 (bad), 2 (poor), 3 (fair), 4 (good) and 5 (excellent).

According to the related research work, QoE estimation approaches are divided into three types: subjective, objective and hybrid solutions. The first one requires participants to evaluate their perceptions about multimedia applications, so it is costly, time consuming and ineffective with real-time monitoring due to human intervention. The second one constructs an objective model estimating user perception using parameters (e.g. network parameters, application parameter, etc.), but identifying an effective model is sometimes complicated. The third one is a hybrid approach that uses a subjective dataset to learn QoE models based on ML algorithms. Consequently, it can combine the advantages of both subjective and objective approaches.

Therefore, in this chapter, we implemented the hybrid approach for QoE estimation, offering opportunities to estimate MOS in real time. The MOS dataset and description about the testbed building it (e.g. number of participants, understanding level of participants, etc.) are published to Github (Amour et al. 2015a). We have chosen the *random forest* algorithm to estimate the MOS score because it has less root mean square error than other ML approaches (Amour et al. 2019). MOS is estimated according to network parameters (e.g. latency, packet loss, etc.). The detail of QoE estimation is described in Amour et al. (2019).

2.3.3.4. *Other parameters*

For other parameters, link utilization, the number of packets sent, the number of packets received, the number of bytes sent, the number of bytes received and the number of flow entries in switch are calculated according to the OpenFlow-based monitoring approach via the *Portstatistic* application programming interface (API) (Team 2021) in SDN controllers.

2.3.3.5. *Measurement frequency*

In parameter measurement, a frequency of polling for measurement is an essential factor. There are two methods for identifying the frequency of polling, including periodic polling and event-triggered polling (Lee et al. 2014). The former requests a measurement periodically after a given time, while the latter requests the measurement when a pre-defined event occurs. The event-triggered polling requires building a database about pre-defined events, and it is not effective with unnoticed events. Periodic polling can overcome this obstacle and offer opportunities to monitor the network in real-time effectively. Therefore, we implement periodic polling in the parameter measurement module. The frequency is set to three seconds (standard window time), as in existing work (Kim et al. 2016; Phan and Fukuda 2017). According to the parameter measurement module, we can collect data to build troubleshooting datasets for network troubleshooting. These datasets are described in the following sections.

2.4. Troubleshooting dataset

2.4.1. *Datasets for root cause analysis*

There are several troubleshooting datasets related to troubleshooting tickets, network parameters in network function virtualization (NFV), etc. Table 2.2 describes the existing troubleshooting dataset. However, there is no troubleshooting dataset about network parameters for identifying root causes of anomalies (e.g. link failure, switch failure, etc.).

Therefore, in this chapter, we build and contribute the troubleshooting dataset for network troubleshooting. Recently, during the Covid-19 pandemic, the network infrastructures of NOs have had to suffer high pressure from a vast amount of Internet traffic. This results from many network devices accessing the Internet for education, remote working, entertainment, etc. Consequently, this leads to congestion in the network. According to a report of NOs, congestion is one of the popular issues in the network (Zeng et al. 2012b). Therefore, we take congestion into account to generate anomalies in the network and construct the troubleshooting dataset related to three root causes of congestion containing link failure, switch failure and buffer overload. It is not easy to obtain a dataset related to these root causes because it happens unpredictably in the network. Therefore, many studies

(Cid-Fuentes et al. 2018; Hong et al. 2020) use a fault injection technique to generate root causes. Link failure is simulated by adding latency to links, while switch failure is simulated by generating rule failures to the switches to obtain packet loss. Buffer overload is simulated by sending a huge amount of traffic exceeding the link's capacity. These root causes are simulated in two scenarios including static and dynamic networks. In the static network, the delay and loss in the links are fixed with a specific value. In the dynamic network, the link state is changed between normal and error state according to the Gilbert–Elliot (GE) model (Bildea et al. 2015), following a probability value after a given time (400 s). In this model, the probability of changing from normal state to error state is *p,* while the figure for changing from error state to normal state is *r*. When a link is in the error state, the loss and delay will be modified as described in Table 2.3.

Data types	Dataset	Class	Number of samples
Troubleshooting ticket	Servicenow (Kishore 2021)	Network, user maintenance, database, application workbench, and security	3,001
	Github Troubleshooting Dataset (Github 2021)	Github bug, enhancement and question	30,000
Network parameter	NFV Dataset (2021)	Normal, CPU bottleneck, memory bottleneck, and I/O bottleneck	3,693

Table 2.2. *Existing troubleshooting dataset*

In each network condition, we first extract nine features corresponding to network and standard parameters because of existing studies (Li et al. 2018; Amour et al. 2019) and API *PortStatistics* (Team 2021) in the controller. These parameters contain latency, packet loss, link utilization, number of packets sent, number of packets received, number of bytes sent, number of bytes received, number of flow entries in switch and QoE. Then, these parameters on each link will be aggregated to calculate the paths' features to identify the root cause of anomalies. After that, these features are normalized into a value between 0 and 1 according to max normalization (Singh and

Singh 2020). In RCA, we consider time-series features and sample aggregate features of 10 consecutive time steps to identify the root causes. We built two datasets for network troubleshooting (one for each network condition). The number of network states corresponding to each root cause is depicted in Table 2.4. These datasets are published in Tong (2021c).

Root Causes	Static Network	Dynamic Network
Link failure	Delay: 125ms	Delay: [25, 50, 75, 100, 125ms]
Switch failure	Loss: 50%	Loss: [10, 20, 30, 40, 50%]
Buffer overload	Bandwidth: 10 mbps, Sending rate: 60-100 mbps	Bandwidth: 10 mbps, Sending rate: 60-100 mbps

Table 2.3. *Considered network conditions*

Root Causes	Static Network	Dynamic Network
Buffer overload	6900	6900
Link failure	9995	19710
Switch failure	7560	17565

Table 2.4. *Datasets for root cause analysis.*

2.4.2. Dataset for traffic classification

To validate the performance of the traffic classification module, we built a traffic classification dataset for QUIC traffic over three weeks (starting from March 2018). To capture the network traffic of video streaming, chat, VoIP and Google play music applications, we used the Selenium WebDriver (Selenium 2021) in Google Chrome running on Ubuntu 16.04 OS. We also used Quic-go (Clemente 2021) to transfer data between servers and clients using the QUIC protocol. Then, network traffic of file transfer application is captured. We captured approximately 150 GB of network traffic, including over 20,000 network flows of five QUIC-based applications. The detail is described in Table 2.5. This dataset is published in Tong (2021a).

Application	Number of samples (flows)
Chat	2,783
VoIP	2,608
File transfer	4,451
Video streaming	5,844
Google play music	4,349

Table 2.5. *Classification dataset*

In the following chapter, we present in detail a novel encrypted traffic classification approach for QUIC traffic. The approach is to identify the application class in the context of encrypted traffic.

2.5. Conclusion

In this chapter, we presented a novel troubleshooting architecture adapted to the context of encrypted traffic. We also showed the proof of concept of this architecture in SDN environments. After that, we presented a parameter measurement module to collect network parameters composed of latency, packet loss, link utilization, number of packets sent, number of packets received, number of bytes sent, number of bytes received, number of flow entries in switch and QoE. The purpose is to publicly provide troubleshooting datasets for the research community.

In addition to the network parameters, we identified the application class to facilitate the application-aware remediation approaches in network troubleshooting.

3

Traffic Classification: Novel QUIC Traffic Classifier Based on Convolutional Neural Network

"Predicting the future isn't magic, it's artificial intelligence"

Dave Waters

In Chapter 3, we present a traffic classification module to identify the application class for QUIC traffic. The application class also plays an important role in application-aware remediation approaches in network troubleshooting. This module is briefly described in the novel troubleshooting framework (Figure 2.2).

3.1. Introduction

Traffic classification (Lopez-Martin et al. 2017) plays an essential role in network troubleshooting for NOs. The objective of traffic classification is to identify application classes in the network to implement application-aware mechanisms to address network problems and meet strict SLA requirements. Concretely, traffic classification is used widely in application-aware routing, billing policies, intrusion detection systems, identifying security threats, enforcing QoS requirements, etc. (Rezaei et al. 2019). Therefore, traffic classification is studied thoroughly by the research community.

In the past, many attempts to identify the application class relied on port-based approaches and payload-based approaches. The first approach uses port numbers in TCP/UDP packets to identify the application class. However, it is not effective because modern applications and protocols are not always tied to a specific port. Besides, many networks change the port number in the packets using network address translation (NAT), leading to inaccurate results for traffic classification. The second approach (Adami et al. 2015) inspects a payload of packet to identify its signatures corresponding to a particular application, called Deep Packet Inspection. This approach can achieve high accuracy, but it is limited due to encrypted network traffic. Concretely, network traffic nowadays is encrypted to protect data and user's privacy, so the signature of application in the payload of packet is hidden.

To overcome drawbacks of such approaches, the statistic-based approach using machine learning (ML) (Lopez-Martin et al. 2017) emerges. This approach analyzes high-level information obtained from network flows (e.g. TCP window size, packets inter-arrival times, etc.) using ML to identify the application class. Flow-based features can be extracted from all packets of a flow (Fan and Liu 2017; Fathi-Kazerooni et al. 2019) or the first few packets of a flow (Lopez-Martin et al. 2017; Salman et al. 2018). However, these approaches require us to identify predefined flow-based features with human intervention, so it requires a knowledge base of considered applications and a large analysis time. Consequently, many studies focus on a traffic classification approach using deep learning (DL). The objective of the DL algorithm is to combine feature extraction and classification modules into one module, so this approach does not require human intervention for the feature extractions. This method converts the payload of packet into bytes to create a vector representing an application class. Then, DL algorithms analyze this vector to automatically extract packet-based features (implicit features) and classify these features into the application classes for HTTPS (Wang et al. 2018) and VPN traffic (Wang et al. 2017; Lotfollahi et al. 2020).

Similar to HTTPS, Google has recently developed Quick UDP Internet Connection (QUIC) (Langley et al. 2017), a new transport layer network protocol on the top of UDP. QUIC contains many advantages related to connection establishment, congestion control, multiplexing without head of line blocking and connection migration. Besides, QUIC also provides a security protection equivalent to Transport Layer Security/Secure Sockets Layer (TLS/SSL).

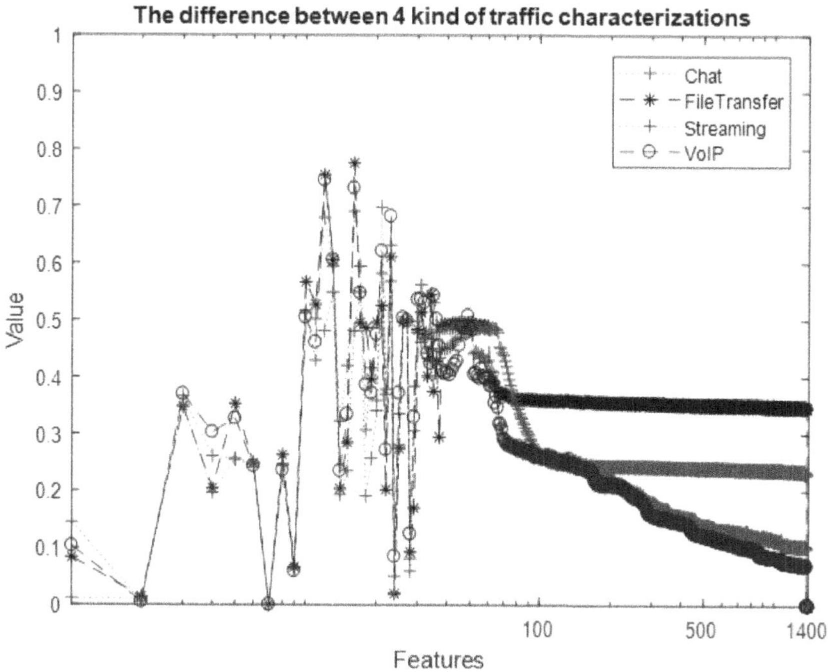

Figure 3.1. *Byte in payload of QUIC packets for different applications. For a color version of this figure, see www.iste.co.uk/tong/troubleshooting.zip*

Many existing studies focus on traffic classification solutions for HTTPS (Wang et al. 2018) and VPN traffic (Wang et al. 2017; Lotfollahi et al. 2020). In this book, we concentrate on a traffic classification solution for QUIC traffic. From the experiments, the investigation on bytes in payload of the QUIC packets (Figure 3.1) illustrates that the DL-based traffic classification approaches cannot be effective for several QUIC-based applications including chat and VoIP applications. The reason is that there is less of a difference in the payload of packets for chat and VoIP applications.

Therefore, in this book, we present a novel traffic classification method using convolutional neural network (CNN). It is a hybrid between the statistical-based method and the DL-based method. The proposal contains two main stages. The first one uses flow-based features and an ML algorithm to classify the network traffic into chat, VoIP and other applications corresponding to elephant flows (large continuous flows in total bytes). The second one takes into account packet-based features and CNN to classify the elephant flows into video streaming, file transfer and Google Play Music. Chat, VoIP, video streaming, file transfer and Google Play Music are taken into account here because it is reported that these applications will comprise over 80% of global IP traffic by 2022 (Cisco 2021b).

The remainder of this chapter is as follows. Section 3.2 presents background information about CNN and the characteristics of QUIC-based applications. Related work on traffic classification is introduced in section 3.3. Section 3.4 describes the novel traffic classification approach for QUIC traffic. Section 3.5 shows experimental results. Section 3.6 presents a conclusion, highlighting future prospects.

3.2. Background

3.2.1. *Convolutional network*

Convolutional network (O'Shea and Nash 2015; Goodfellow et al. 2016), known as convolutional neural network (CNN), is a specialized artificial neural network for data processing. It can be applied to pattern recognition, DGA botnet detection, traffic classification, etc. (Lopez-Martin et al. 2017). artificial neural network (ANN) (Goodfellow et al. 2016) is a kind of ML algorithm inspired by the biological neural networks. However, ANN contains two main drawbacks (Goodfellow et al. 2016) in the context of a large input size. ANN has connections between each neuron of a layer to each neuron of the following layers. It implements a matrix multiplication by a matrix of separate parameters. Besides, each element of weight matrix in ANN is utilized only once to compute the output of the next layers.

CNN contains characteristics including spare connectivity, parameter sharing and equivariant representations (Goodfellow et al. 2016). Consequently, it offers opportunities to overcome the disadvantages of ANN. Spare connectivity means that each neuron in a layer can interact indirectly with a large portion of the input to make a kernel smaller. With the parameter sharing, each kernel element is utilized at every position of the input to use the same parameters for more than one time. The third characteristic indicates that a translation of input features results in an equivalent translation of outputs. These characteristics result in a reduction of memory requirements and processing time to obtain better performance compared to ANN.

3.2.2. *Characteristics of QUIC-based applications*

Figure 3.2 shows the differences between QUIC-based applications related to flow-based features. In this chapter, we use a collected dataset (Tong 2021a) with five applications containing VoIP (VC), Chat (C), Video Streaming (VS), Google Play Music (GPM) and File Transfer (FT).

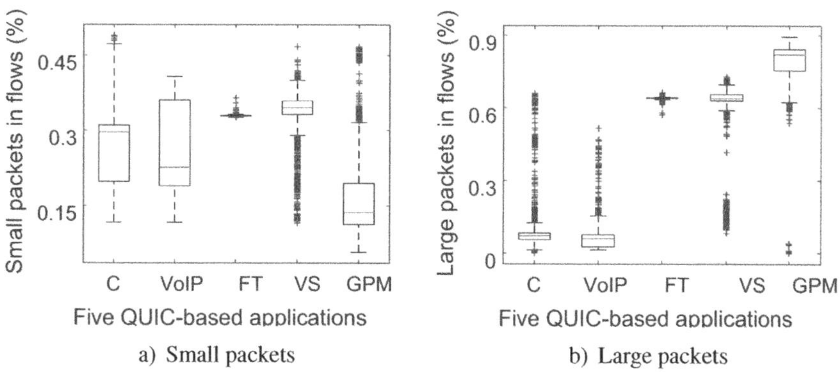

Figure 3.2. *Percentage of small and large packets in flows. For a color version of this figure, see www.iste.co.uk/tong/troubleshooting.zip*

In QUIC-based applications, QUIC is responsible for establishing and finishing connections and transferring data between end-users. There are significant differences between these applications. In the chat application of Google Hangout, all packets transferred between end-users are supported by QUIC, but the data transferred are insufficient. Therefore, most packets in flows are small packets whose packet lengths are less than 150 bytes. It is similar to VoIP of Google Hangout. In Figure 3.2a, a range of small packets in the flows of chat and VoIP is larger than the figure for others.

In file transfer, video streaming and Google Play Music, servers and clients frequently transfer a large amount of data, so there is a high number of large packets whose packet lengths are bigger than 1,000 bytes in the network flows. In Figure 3.2(b), the percentage of large packets in the flows of these applications is over 60%. Moreover, the clients frequently reply small packets to the servers, so the percentage of small packets of these services is approximately 30%, except for Google Play Music (approximately 15%). However, the range of percentage of small packets in these applications is slightly tight.

3.3. Traffic classification approaches

In this section, we present related work on traffic classification consisting of port-based approaches, payload-based approaches, statistic-based approaches and DL-based methods.

3.3.1. *Port-based approaches*

Traditionally, the application class is identified according to a well-known port number in a packet. In the past, a particular application is registered to a specific port number in Internet Assigned Numbers Authority (IANA) (June 2021). The port-based approaches analyze the packet to obtain the port number and identify the corresponding application class. Table 3.1 indicates ports registered by IANA and corresponding applications. For instance, SSH applications use port 22, and HTTP applications use port 80 for data transmission.

Registered ports	Applications
20	FTP data
21	FTP control
22	SSH
23	TELNET
25	SMTP
53	DNS
80	HTTP
161	SNMP
443	HTTPS

Table 3.1. *Registered port numbers by IANA for several applications*

This approach is no longer available due to inaccuracy of classification results (Lopez-Martin et al. 2017). In fact, many applications use port numbers from a well-known application (e.g. port 443, etc.). Besides, many network administrators use NAT to change the port number of packets in the network.

3.3.2. *Payload-based approaches*

An alternative for the port-based approach is the payload-based approach, called Deep Packet Inspection. This approach analyzes the payload of packets to obtain the application's signature and identify a corresponding application. Table 3.2 shows the signatures for several peer-to-peer (P2P) applications (Karagiannis et al. 2004).

Application	Signatures	Protocol
eDonkey 2000	0xe319010000, 0xe53f010000	TCP, UDP
BitTorrent	"0x13Bit"	TCP
Ares	GET hash:, Get sha1:	TCP

Table 3.2. *Signatures for several P2P applications*

Adami et al. (2015) identify the VoIP traffic according to the signatures of SIP protocol (e.g. content-type: "application/sdp", etc.). This approach can achieve good performance, but it contains three main drawbacks. Firstly, searching the signatures in the payload of packets requires high resource consumption in the network devices. Secondly, the packet's payload is hidden due to encrypted traffic nowadays, so this leads to many obstacles for signature extraction. Finally, this approach requires updating the signature frequently for a new version of applications or new applications.

3.3.3. *Statistic-based approaches*

The emergence of statistic-based approaches using ML is a potential solution to overcome the limitations of port and payload-based approaches. This method extracts flow-based features corresponding to network flows and analyzes it using the ML technique to classify the network flows into different applications.

Williams et al. (2006) proposed a traffic classification using a statistical-based approach. This method extracts the flow-based features from all packets of flows and analyzes it using ML algorithms (e.g. Naive Bayes, C4.5, Bayesian Network, etc.) to classify network traffic into FTP, Telnet, SMTP, DNS and HTTP. The flow-based features contain a protocol, flow duration, flow volume in bytes, packet length and inter-arrival time in the flows. Similarly, Fathi-Kazerooni et al. (2019) proposed a traffic classification approach using the random forest (RF) algorithm. This approach collects 10 flow-based features from all packets of flows and analyzes it to identify six applications including Chrome, Google Drive, One Drive, One Note, Spotify and WhatsApp. However, these approaches are not able to early detect applications because these approaches extract and analyze flow-based features after the flows terminate.

Many studies focused on statistical-based approaches for early application detection. Amaral et al. (2016) proposed a novel traffic classification approach for the SDN environment. The objective is to collect 12 features from the first five packets of flows (e.g. packet size, packet time-stamp, inter-arrival time, etc.) and analyze it using ML algorithms (RF, stochastic gradient boosting and extreme gradient boosting) to identify eight applications. Besides, Lopez-Martin et al. (2017) proposed a traffic classification approach for IoT network. This approach extracts six time-series features of the first 20 packets of flows and analyzes it using DL algorithms (RNN, CNN and CNN+RNN) to classify network traffic into 108 applications. However, this approach uses the port number in traffic classification, so this can result in uncertain results when many applications use the same ports (e.g. port 443, etc.). Moreover, the statistic-based approaches require us to identify predefined flow-based features with human intervention. Consequently, these approaches require a knowledge base of applications.

3.3.4. *DL-based approaches*

The DL-based approach is a variant of the statistic-based approach. The main goal of the DL algorithm is to combine the feature extraction and classification modules into one module. This approach extracts packet-based features (implicit features) according to analyzing the packet's payload using the DL algorithm, unlike the statistic-based approach which extracts the predefined flow-based features. Then, the packet-based features are analyzed according to a classifier in the DL algorithm to identify the application class.

Lotfollahi et al. (2020) proposed “Deep Packet”, a novel traffic classification using DL algorithms. Deep Packet processes 1,500 bytes in the payload of packet to obtain a vector. Next, this vector is analyzed in the DL algorithms to classify network traffic into email, chat, FTP, Skype, torrent, etc. In this approach, the authors evaluate the performance of two DL algorithms including stacked autoencoder and CNN. However, using all bytes in the packet’s header (e.g. port number, etc.) can lead to uncertain results because many applications use the same ports. Similarly, Wang et al. (2017) analyze the first 784 bytes in each packet using CNN to identify six VPN and six non-VPN applications. Moreover, Wang et al. (2018) proposed an encrypted traffic approach using the DL algorithm in the home gateway. This method processes the first 1,480 bytes of each packet and analyzes it using ML algorithms to identify 15 applications in the context of encrypted traffic (e.g. Facebook, Gmail, SFTP, etc.). To select an appropriate algorithm, the authors evaluate the performance of multilayer perceptron (MLP), stacked autoencoder and CNN. However, these DL-based approaches only take into account the payload of a packet in network flows to identify the application class. To improve the performance of traffic classification, Lim et al. (2019) have proposed a traffic classification using multi-layer long short-term memory (LSTM), which considers payload of many packets in a network flow. This approach analyzes the payload of the first 30 packets in a flow to create a three-dimensional vector and also analyzes it using multi-layer LSTM to identify eight applications.

The existing studies focus on traffic classification approaches for non-encrypted traffic or encrypted traffic including HTTPS and VPN traffic. Therefore, in this book, we focus on a traffic classification approach for QUIC traffic. After investigating the characteristics of QUIC-based applications, we realized that DL-based approaches are not effective for several applications (e.g. chat and VoIP, etc.). Consequently, we present a hybrid QUIC traffic classification approach using both flow-based features, packet-based features and DL algorithms.

3.4. Novel traffic classification method for QUIC traffic

In this book, we present a new traffic classification method using flow-based features, packet-based features and CNN. The detail is described as in Figure 3.3(a). There are two kinds of network flows: mice flows (small continuous flows in total bytes) and elephant flows (large continuous flows in total bytes). Therefore, the proposal comprises two main stages of classification. In the first stage, flow-based features are analyzed in ML algorithms (e.g. random forest (RF), support vector machine (SVM), etc.) to identify chat and VoIP applications (mice flows). After that, packet-based features of elephant flows are extracted and analyzed according to CNN to classify elephant flows into file transfer, video streaming or Google Play Music.

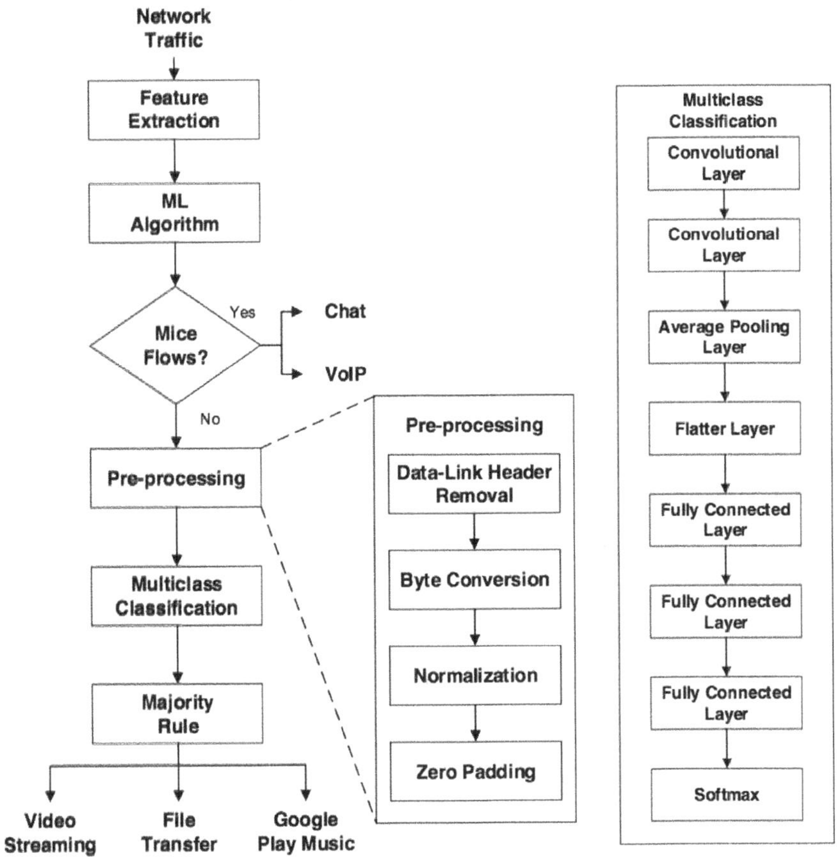

a) Overall structure

b) Multiclass classification module using CNN

Figure 3.3. *Novel traffic classification approach for QUIC traffic*

3.4.1. *Traffic collection*

In traffic classification, we need to collect network traffic and analyze it to identify the application classes. Network traffic can be collected directly from OpenFlow switches in the SDN environment, but it is not effective with a tremendous amount of network traffic. Consequently, it is necessary for a network monitoring standard supporting sampling technique to reduce the volume of traffic. There are two sampling techniques including flow sampling and packet sampling. Flow sampling (e.g. NetFlow, etc.) samples a flow with a specific probability, aggregates all packets of this flow into flow records and exports them to collectors. Packet sampling (e.g. sFlow, etc.) samples each packet with a specific probability and aggregates sampled packets into sFlow records and exports them to collectors. The flow sampling technique can collect too many packets of several flows and no packets for other flows. Therefore, it is ineffective for traffic classification. We can partly collect packets of all flows in the network with the packet sampling technique. Therefore, sFlow (Ujjan et al. 2020) is used here in the traffic classification approach. sFlow was widely used by vendors including IBM, HP, OpenVswitch, etc.

sFlow contains two main elements including sFlow agents and sFlow collectors. sFlow agents are implemented at ingress nodes (edge nodes) in the network. When there is network traffic traversing the sFlow agents, it will sample these packets, aggregate them and export to the sFlow collectors for the traffic classification.

3.4.2. *Flow-based features*

In the first classification stage, flow-based features are used to identify chat and VoIP applications. According to section 3.2.2, there are differences between five applications, particularly packet length. Therefore, we take into account eight flow-based features in the first classification stage: average payload length; percentage of small, medium and large packets in the network flows in both directions between clients and servers. The length of small, medium and large packets range from 0 to 150, 150 to 1,000 and over 1,000 bytes, respectively.

3.4.3. *Preprocessing*

The pre-processing module processes QUIC packets of elephant flows, converts its payload into bytes and normalizes it into decimal values for further modules. This module contains four main steps consisting of data-link header removal, byte conversion, normalization and zero padding.

The data-link header contains information related to the physical layer, which plays an important role in frame forwarding in the network. However, this information is useless for traffic classification, so the data-link header will be filtered in the data-link header removal step. Besides, we only use the payload of packets because the remaining information in the packets (e.g. port number, etc.) can result in uncertain results in the traffic classification. After that, the payload of packets is converted from bit to byte to reduce an input size. Then, these bytes are normalized by dividing by 255, the maximum value for a byte. CNN requires the same input length while the length of QUIC packets varies from over 50 to approximately 1,400 bytes. Therefore, the packets insert zero values in the zero-padding step to have a similar length for each packet. If the packet length is less than 1,400, these packets are padded with zero values at the end of packets. Finally, each packet is represented by a vector containing 1,400 decimal values.

3.4.4. *Novel traffic classification method*

Figure 3.3(a) describes the new traffic classification method. Firstly, network traffic is collected, and then the first few packets of each flow are processed in the feature extraction module to extract eight flow-based features representing the network flow. The more packets in a flow we use, the higher accuracy we can obtain. However, it results in a high processing time. Therefore, we consider the first 20 packets of each flow to extract flow-based features as in existing studies (Lopez-Martin et al. 2017). Then, these features are analyzed in ML algorithms that classify the network flows into chat, VoIP and elephant flows. ML algorithms are considered as classifiers for identifying the application class. There are many ML algorithms (e.g. RF, SVM, MLP, etc.), but the RF algorithm outperforms the others (Zhou et al. 2020; see section 3.5). Consequently, RF is considered to be an ML algorithm in the first classification stage. After identifying chat and VoIP applications, the remainder of network traffic (elephant flows) is processed in the second classification stage to classify elephant flows into video streaming, file transfer and the Google Play Music application.

Firstly, the packet of elephant flows is processed in the preprocessing module to extract the vector of 1,400 decimal values representing each packet. Then, this vector is analyzed in the multiclass classification module to identify the application class. The structure of this module is described in Figure 3.3(b). The size of input vector is large (1,400 values), so ANN as well as traditional ML algorithms (e.g. SVM, MLP, etc.) are not effective in data processing (Goodfellow et al. 2016). Besides, CNN contains characteristics including spare connectivity, parameter sharing and equivariant representations. This offers opportunities to learn more effective representations in comparison to traditional ML algorithms (Goodfellow et al. 2016). Therefore, CNN is taken into account in the multiclass classification module.

The structure of the multiclass classification module contains five essential layers including a convolutional layer, an average pooling layer, a flatter layer, a fully connected layer and a softmax. The convolutional layer is considered as a feature learning to extract packet-based features representing a packet in elephant flows. The average pooling layer replaces the input vector at a certain location with an average value of nearby values, while the flatter layer flattens an output obtained from the average pooling layer into a one-dimensional vector. In the fully connected layer, each neuron has complete connections to each neuron in the following layers. The softmax layer is a generalization of a logistic function that squashes an n-dimensional vector into an n-dimensional vector with element values between 0 and 1. The combination of the fully connected layer and softmax is similar to a classifier that indicates a relationship between the input vectors and the application classes to classify the elephant flows into different kinds of applications.

In each elephant flow, we extract packet-based features of the first few packets, classify these packets and aggregate classification results to identify application class for the elephant flow according to the majority rule. The majority rule is a decision rule that selects an alternative with high votes. If most packets among these packets are classified as an application, the flow will be assigned to that application. If the vote of applications is equal, a classification probability of these packets is used to identify the application class. The higher the number of packets in each flow classified, the higher the accuracy we can obtain. Nevertheless, this can lead to a high processing time. Therefore, we consider the first 10 packets of each elephant flow to identify the application class. The detail of the majority rule is described in Algorithm 1. $label(id)$ is an application class of flow after the classification and $prob_{ij}$ is a probability of packet $packet_i$ classified as application app_j in the flows. argmax is a function that returns an index of the highest value in the input. m and n are the number of application classes and the number of packets required for the classification in each flow, respectively.

Algorithm 1 Majority rule

Require: $id, FlowId, label, FlowLabel, app, Count, prob, Prob, a$

while id in $FlowId$ **do**
    for $j = 1$ to m **do**
        for $i = 1$ to n **do**
            if $label(id)$ is app_j **then**
                $a_{ij} = 1$
            else
                $a_{ij} = 0$
            end if
        end for

$$Count_j = \sum_{i=1}^{n} a_{ij}$$

$$Prob_j = \sum_{i=1}^{n} prob_{ij}$$

    end for
    if all $Count_j$ unique in $Count$ **then**
        $FlowLabel(id)$ = argmax($Count$)
    else
        $FlowLabel(id)$ = argmax($Prob$)
    end if
    Return application class $FlowLabel(id)$
end while

The proposal contains two main algorithms including RF and CNN. Therefore, time complexity of the proposal is related to time complexity of these algorithms. Regarding the time complexity of the RF algorithm, its time complexity in the training phase is $O(n_O * log(n_O) * d * k)$, where n_O is the number of input sample, d is dimensional of data and k is the number of trees in the forest. The time complexity in testing phase is $O(k * d')$ where d' is the depth of tree (Gao et al. 2018).

As for the time complexity of CNN, total time complexity of all convolutional layers (He and Sun 2015) is estimated by equation [3.1]:

$$O(\sum_{l=1}^{d} n_{l-1} \cdot s_l^2 \cdot n_l \cdot m_l^2). \tag{3.1}$$

where d is the number of convolutional layers, n_{l-1} is the number of input channels of the lth layer, n_l is the number of filters in the lth layer, s_l is the spatial size of the filter and m_l is the spatial size of the output feature map.

This time complexity is applied in both training and testing phase, and training time is approximately three times testing time.

3.5. Experimental results

3.5.1. *Dataset specification*

In the experiment, we implement a testbed to collect network traffic and publically provide a real traffic classification dataset from March to April 2018 (Tong 2021a). Network traffic of video streaming, chat, VoIP and Google Play Music applications are captured by watching videos on YouTube, accessing Google Hangout and Google Play Music automatically according to Selenium WebDriver Selenium (2021) in Google Chrome running on Ubuntu 16.04 OS. Network traffic of file transfer application is collected by transferring data between servers and clients using *quic-go* (Clemente 2021), an open-source library supporting QUIC protocol. As a result, approximately 150 GB of network traffic was collected, including over 20,000 network flows of five QUIC-based applications. The detail is described in Table 3.3. In the first classification stage using the random forest (RF) algorithm, the dataset is divided fivefold. Concretely, the training and testing phase comprise 20% and 80% of the dataset, respectively. In the multiclass classification, the dataset is also divided fivefold. The training, validation and testing comprise 45%, 5% and 50% of the dataset, respectively.

The proposal is written in Python using Keras library (Chollet 2016a) and scikit-learn tools (Pedregosa et al. 2011). Besides, all experiments are implemented in a workstation (Intel(R) Xeon(R) CPU E5-2640 v3 @ 2.6 GHz and 16 GB of RAM) running Ubuntu 16.04.

Application	Number of samples (flows)
Chat	2,783
VoIP	2,608
File transfer	4,451
Video streaming	5,844
Google Play Music	4,349

Table 3.3. *Dataset specification*

3.5.2. *Performance metrics*

In this section, we present measures (Sokolova and Lapalme 2009) to evaluate the performance of the proposal consisting of precision, recall and F1-score (F-measure). Precision is a percentage of relevant flows that are retrieved, while recall is a percentage of relative retrieved flows. F1-score represents a harmonic mean between precision and recall. The detail of these parameters are described in equations [3.2]–[3.4].

$$Precision = \frac{TP}{TP + FP}. \quad [3.2]$$

$$Recall = \frac{TP}{TP + FN}. \quad [3.3]$$

$$F1 - score = \frac{2}{1/Recall + 1/Precision}. \quad [3.4]$$

Quality of overall classification can be evaluated in two ways (Sokolova and Lapalme 2009). In macro-averaging, a metric is averaged over all equally-treated classes. Micro-averaging is based on cumulative True Positive (TP), False Positive (FP), True Negative (TN) and False Negative (FN) of the dataset.

Class	Precision			Recall			F1-score			Nb Samples
	MLP	SVM	RF	MLP	SVM	RF	MLP	SVM	RF	
Chat	0.9800	0.9759	0.9671	0.0072	0.8448	0.9454	0.0143	0.9057	0.9562	2,227
VoIP	0.6298	0.9460	0.9671	0.9335	0.9608	0.9680	0.7529	0.9534	0.9675	2,086
Elephant flows	0.8690	0.9582	0.9757	0.9628	0.9800	0.9797	0.9135	0.9690	0.9777	11,351
Micro	0.8234	0.9587	0.9724	0.8234	0.9587	0.9744	0.8234	0.9587	0.9734	15,664
Macro	0.8263	0.9600	0.9679	0.6278	0.9286	0.9634	0.5536	0.9427	0.9656	

Table 3.4. *Performance metrics of ML algorithms in the first stage of classification*

Algorithm	Average Training Time (ms)	Average Testing Time (ms)
MLP	0.25505	0.00042
SVM	7.76140	0.01472
RF	0.11787	0.03165

Table 3.5. *Time complexity of ML algorithms in the first stage of classification*

3.5.3. *Performance analysis*

This section is dedicated to evaluating the performance of the traffic classification method which contains two stages of classification. In the first stage, we evaluate the performance and time complexity of ML algorithms (e.g. SVM, RF, etc.) to select an appropriate ML algorithm. In the second stage, we evaluate its performance over different scenarios including different subsets of the input vector and various loss functions. We thus evaluate the running time of the traffic classification approach.

3.5.3.1. *First stage of classification*

The performance of MLP (Pedregosa et al. 2021a), SVM (Pedregosa et al. 2021b) and RF (RF 2021) are described in Table 3.4. The micro and macro precision, recall and F1-score of MLP are the lowest in three ML algorithms with below 83%, while the figures for RF are the highest with approximately 97%. Table 3.5 indicates the time complexity of ML algorithms in the first stage of classification. We calculate the training time and testing time for the training and testing dataset and then generate the average training and testing time for a sample in these datasets. Although MLP does not have good performance, its training and testing time are the lowest in three algorithms with 0.25505 and 0.00042 ms, respectively. The average testing time of SVM is lower than the figure for RF with 0.01472 and 0.03165 ms, respectively. However, the difference between macro recall of SVM and RF is approximately 4%. Therefore, RF is selected as an ML algorithm in the first stage of classification.

With RF algorithm, the precision, recall and F1-score of chat and VoIP applications are impressive, with approximately 96%, except for the recall of chat application which is 94.54%. Chat and VoIP applications have similar value regions in flow-based features, so several flows of VoIP application are classified incorrectly to chat application. This leads to lower results for the chat application. In particular, the performance of elephant flows (file transfer, video streaming and Google Play Music) are over 97%. The reason for this is that there is a difference in flow-based features between chat, VoIP and elephant flows. As a result, the micro- and macro-averaging precision, recall and F1-score of the dataset achieve over 96%.

3.5.3.2. *Second stage of classification*

After classifying the network traffic into chat, VoIP or elephant flows, each packet of elephant flows is preprocessed in the preprocessing module to obtain a vector of 1,400 decimal values, and this vector is analyzed in the second stage of classification to classify into video streaming, file transfer or Google Play Music.

In the first scenario, we investigate the influence of different subsets of input vectors in the second stage of classification. We implement the experiments on five subsets with the sparse_categorical_crossentropy loss function to select the subset with good performance. Figure 3.4 indicates the macro-averaging precision, recall and F1-score of five subsets including the first 300, 600, 900, 1,200 and 1,400 decimal values. There is an upward trend in the macro-averaging precision, recall and F1-score in five subsets. The performance for the first 300 and 900 decimal values are not good, with over 70% for the macro-averaging F1-score. The reason for this is that the precision of video streaming application with the first 300 values is 64.34%. Besides, flows of video streaming are classified incorrectly to the flows of file transfer, so this reduces the recall of file transfer in the experiment of the first 300 values. It is similar to the subset of the first 900 values. The discriminant power with the subset of 1,400 values is larger than the figure for others, so the macro-averaging precision, recall and F1-score are the highest, with approximately 99%.

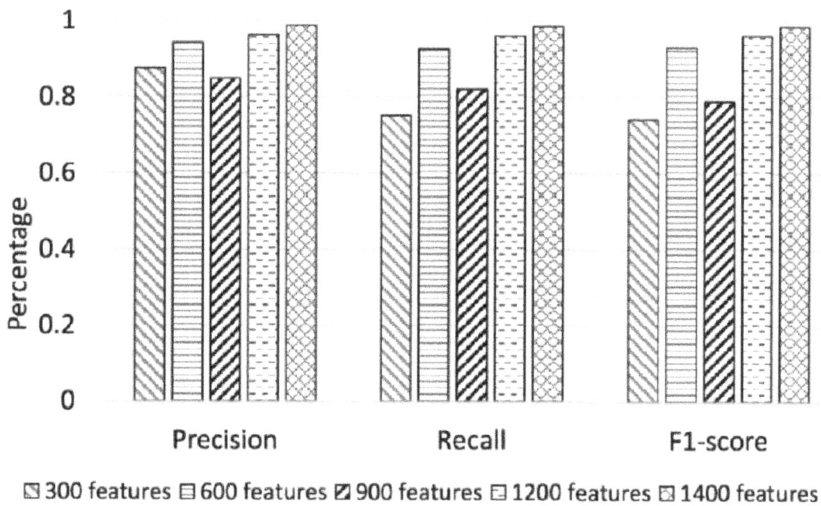

Figure 3.4. *Macro-averaging precision, macro-averaging recall and macro-averaging F1-score in different subsets. For a color version of this figure, see www.iste.co.uk/tong/troubleshooting.zip*

In the second scenario, we investigate the influence of loss functions (Chollet 2016b) on the convolutional neural network (Figure 3.5). As discussed above, the performance of the dataset with 1,400 values is the highest in five subsets, so we will use this dataset in the second scenario. In this scenario, we evaluate the performance of three kinds of the loss function (categorical_hinge (Hinge), Mean_Squared_Error (MSE) and Sparse_Categorical_Crossentropy (SCCE)) to select the appropriate loss function. Hinge loss is the loss function that is notably used in support vector machine for "maximum-margin" classification. In Figure 3.5, the performance of hinge loss is not good because the macro-averaging F1-score is over 20%. The reason is that many flows of file transfer are classified incorrectly. Mean squared error is the loss function that measures an average square of error. The performance of MSE loss is higher than the figure for hinge loss. A noticeable feature is that the performance of SCCE is the highest in three loss functions. The macro-averaging of SCCE is the largest, with approximately 99%.

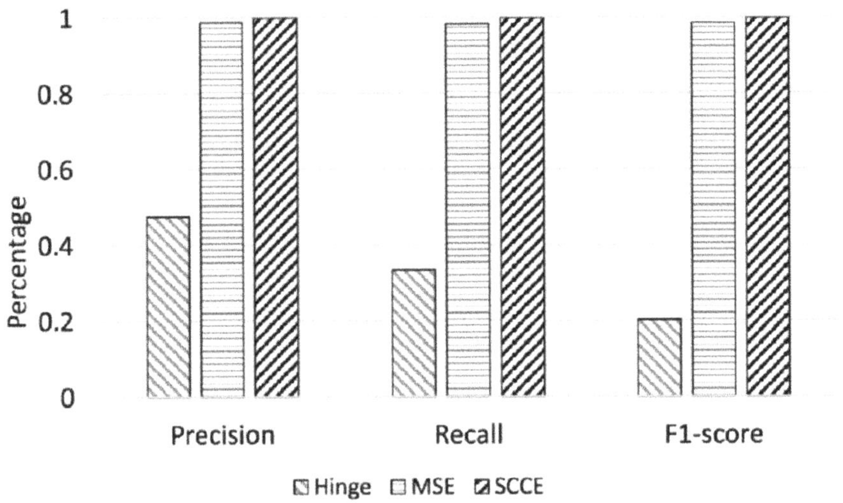

Figure 3.5. *Macro-averaging precision, macro-averaging recall and macro-averaging f1-score in different loss functions. For a color version of this figure, see www.iste.co.uk/tong/troubleshooting.zip*

Briefly, we consider the dataset with 1,400 values corresponding to 1,400 bytes in the payload of packets and the SCCE loss function in the second stage of classification. Figure 3.6 indicates the overall results of the traffic classification approach with these selections. The micro- and macro-averaging of precision, recall and F1-score are over 96%. These results are summarized according to the results of the first and second stage of classification.

3.5.3.3. *Time analysis*

Regarding the time complexity, the average training and testing time of RF in the first stage of classification are 0.11787 and 0.03165 ms, so this stage can process up to over 33,000 network flows per second. The average training and testing time of CNN in the second stage of classification are 11.1842 and 3.4 ms, so this stage can process approximately 300 network flows per second. Moreover, it is reported that OpenFlow switches can handle from 150 to 750 new flows per second (Taylor et al. 2016). Similarly, Aliyu et al. (2017) reported that HP ProCurve 5406zl with K.15.10.0009 firmware could handle 275 new flows per second. Therefore, the traffic classification approach can operate effectively with SDN switches.

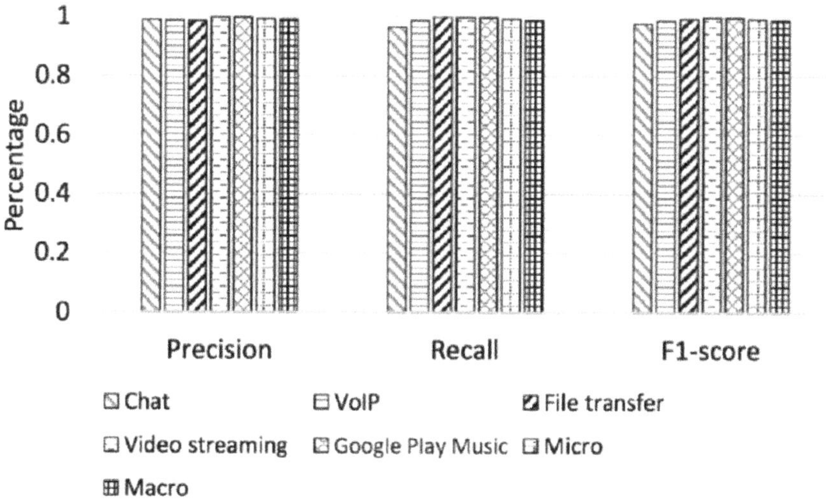

Figure 3.6. *Precision, recall and F1-score of the traffic classification method on five QUIC-based applications. For a color version of this figure, see www.iste.co.uk/tong/troubleshooting.zip*

3.6. Conclusion

In the chapter, we present the novel traffic classification method using the convolutional neural network, flow-based features and packet-based features to identify various QUIC-based applications. There are two main classification stages in this method. The first one is to classify the network traffic into chat, VoIP or elephant flows, whereas the second one aims to classify the elephant flows into video streaming, file transfer or Google Play Music. The experimental results demonstrate that this method can detect five QUIC-based applications with high accuracy (over 96%).

Traffic classification aims to identify application classes for network management (e.g. application-aware traffic engineering, network slicing, etc.). Therefore, time complexity is an essential factor in traffic classification. In the future, we will develop the traffic classification approach with more suitable time complexity. We will thus construct a larger dataset, deeply investigate the flow-based features and compare the performance of this method with existing traffic classification approaches to make it more reliable.

After collecting network parameters from parameter measurement and application class from the traffic classification module, we need to analyze it to identify network anomalies. The objective is to identify when problems occur in the network and make appropriate decisions to optimize network performance. Consequently, in Chapter 3, we present an anomaly detection approach using ML algorithm.

4

Anomaly Detection

> *"The computer was born to solve problems that did not exist before"*
>
> Bill Gates

Chapter 4 presents an anomaly detection approach to identify network anomalies. This approach is implemented in the anomaly detection module, which is depicted in Figure 2.2.

4.1. Introduction

An anomaly is an observation that deviates from other observations to raise a suspicion that it has been generated by a different mechanism (Ahmed et al. 2016). According to its nature, the anomaly can be categorized into point, collective and contextual anomalies (Fernandes et al. 2019). A point anomaly is a deviation of a data sample from usual behaviors. This kind of anomaly is the simplest and is widely studied by the research community. For example, packet loss in the network every day is less than 1%, but it increases to more than 20% on a specific day. This situation is considered as a point anomaly. A collective anomaly happens when a collection of data samples behaves anomalously with an entire dataset. Anomalous behavior is not considered as a collective anomaly, but point anomaly happening continuously in duration is considered as a collective anomaly. A contextual anomaly is an event or behavior considered as an anomaly depending on the context. In contextual anomalies, there are contextual features (e.g. geographic coordinates in spatial data, time in time-series data, etc.) and non-contextual features (e.g. an indicator determining the context of the anomaly, etc.). For example, the throughput of network systems increases significantly at 12 am every day, due to regular backup. This activity creates an outlier related to traffic volume, but it is not considered as an anomaly. However, an outlier at 1 am can be considered as a contextual anomaly.

Anomaly detection plays an important role in the network. It helps NOs to detect problems in the network early on to reduce negative impacts of problems quickly (e.g. high loss, low latency, etc.). Therefore, there have been many anomaly detection techniques produced over the past 10 years including knowledge-based mechanisms, rule inductions, ML-based mechanisms, etc. Despite the many techniques available, the following are obstacles in anomaly detection (Ahmed et al. 2016):

– The boundary between normality and the anomaly is not clear. There is no universal rule to decide whether a behavior is normal or anomalous, except a subjective judgment. Consequently, data can contain noise, which leads to low performance.

– There is a lack of a public labeled dataset in the anomaly detection.

– Designing a general anomaly detection technique is not accessible because there are specific requirements for each context. For instance, an anomaly detection technique for wired networks may have limited usage in the wireless network.

4.2. Anomaly detection approaches

Mdini (2019) grouped the studies on anomaly detection into four main mechanisms, including knowledge-based mechanisms, rule inductions, information theory and ML-based mechanisms. In the following section, we present a brief explanation of these mechanisms and highlight their advantages and drawbacks.

4.2.1. *Knowledge-based mechanisms*

This mechanism (called an expert system) requires human knowledge about anomalies to obtain its patterns, and encodes these patterns in a method that is realized by the machine. Consequently, the knowledge-based mechanism contains two main steps. First, network experts analyze a large amount of data manually to identify anomaly patterns. Then, these patterns are deployed in the network systems to detect anomalies in the future. The knowledge expert can be structured in different ways:

– Rule-based systems: these contain *if-then* rules related to events or alarms. They require an assumption that the same network conditions will result in the same consequences.

– Statistic-based systems: these are an advanced version of rule-based systems. The objective is to create rules based on statistical calculation. Casas et al. (2016) proposed mPlane, a distributed measurement platform for ISP networks. mPlane detects anomalies based on the similarity level between the distribution of network parameters and baseline. The baseline value is the value of network parameters when there is no performance problem in the network. The similarity level between distributions is calculated on Kullback–Leibler divergence. Cid-Fuentes et al. (2018) proposed an anomaly detection approach in distributed systems. This approach detects anomalies using a statistical model. If the radial basis function of an instance is bigger than a threshold, this instance will be considered as an anomaly. Similarly, Chen et al. (2019) proposed a matrix differential decomposition-based anomaly detection and localization in network function virtualization (NFV). This method identifies the anomalies in NFV environments according to the deviation between measured and reference RTT matrix.

The knowledge-based mechanisms suffer from low false positives (FPs) because they only detect anomalies specified by network experts. However, these are two main problems in the approach. First, it is costly and time consuming because the list of anomaly patterns should be exhaustive. Second, it is not effective with new kinds of anomalies that are not defined in the rules. Therefore, the research community has moved toward other anomaly detection approaches.

4.2.2. *Rule inductions*

This approach uses a learning algorithm to learn rules in a deterministic way for anomaly detection. In this approach, there are two main phases including training and testing. In the training phase, this method learns the rules from a large amount of labeled data. Then, the testing phase detects the anomalies according to these rules. There are many learning algorithms for rule induction including decision tree, association rules, etc. The decision tree is a learning algorithm that uses a tree-like decision model for anomaly detection. Association rules generate the rules based on a support and a confidence. Böhmer and Rinderle-Ma (2020) proposed a novel anomaly detection approach based on association rule mining to reduce FPs in anomaly detection. However, it is not effective with high-dimensional data due to high processing time and accuracy issues.

4.2.3. *Information theory*

This approach is related to an assumption that anomalies in a dataset vary its information content. This approach is based on measuring an information indicator of data flows to build an appropriate anomaly detection model. According to related work (Ahmed et al. 2016), there are many measuring functions, including entropy, conditional entropy, relative entropy, relative conditional entropy and information gain. For example, an anomaly detection model needs to look for features with high information gain. If all features have low information gain, performance of anomaly detection is not good. In other words, the higher the information gain of features, the better the performance. This approach is used to obtain an appropriate classifier for anomaly detection, but it is sensitive to noise.

4.2.4. *ML-based mechanisms*

This approach analyzes data points using ML algorithms to classify them into anomaly or normality. In this approach, there are three main steps. First, we identify features representing a data point. Then, we analyze the training data using ML algorithms to learn a model for the classification. Finally, we use this model to classify new data. There are many ML algorithms including decision trees, support vector machines (SVM), neural networks, etc.

In fact, many studies have used decision trees in anomaly detection (Ingre et al. 2017; Khraisat et al. 2018). This approach detects anomalies using a tree-like decision model. The time complexity of this approach depends on the time complexity of algorithms used to construct the trees (e.g. ID3, C4.5, etc.). Although this approach can detect anomalies effectively, it contains two main drawbacks: high dimensional data and the over-fitting problem. Consequently, the random forest (RF) algorithm is studied in anomaly detection by the research community (Primartha and Tama 2017). Random forest builds multiple decision trees on a different subset of the input dataset, and aggregates the output of these decision trees to make the final decision. In this case, it can overcome the over-fitting problem. In the random forest algorithm, a balance between accuracy and computational resources needs to be considered thoroughly. The higher the number of decision trees in the random forest algorithm, the more accurate the classification results and the higher the computational resources.

Bayesian networks are also studied in anomaly detection (Jabbar et al. 2017; Vargas-Muñoz et al. 2018). This approach can be easily and quickly implemented, but it requires an independent assumption between data features. This property is not always guaranteed in the high dimensional data where data features have a correlation. When the independence assumption is not guaranteed, it can lead to low performance.

SVM is another algorithm which is taken into account in anomaly detection (Wang et al. 2017; Safaldin et al. 2021; Hadem et al. 2021). Its objective is to find a hyperplane in data space to detect the anomaly based on a kernel function. However, with non-kernel functions, SVM can suffer from the over-fitting problem, leading to low performance in anomaly detection.

Despite good performance, the ML-based mechanisms contain two main disadvantages. First, they require a labeled dataset to train an anomaly detection model. Second, they cannot effectively detect unknown problems that are not in the training dataset.

4.3. Anomaly detection approach using machine learning

Although there are many anomaly detection approaches, we implement an anomaly detection approach using machine learning in this book. The reason for this is that it offers opportunities to expand to detect anomalies caused by unknown problems easily. This can be achieved by increasing the number of features and expanding the training dataset. Regarding the first drawback of this approach, we simulate root causes of anomalies to collect the training dataset. As for the second drawback, we can frequently update the anomaly detection model when there are new problems.

In this book, we take into account the anomaly detection mechanism in the SDN environment because it offers global visibility and monitoring tools to facilitate anomaly detection (as discussed in Chapter 2). The overall architecture of the ML-based anomaly detection mechanism in the SDN environment is depicted in Figure 4.1. First, the network traffic from the infrastructure layer is collected and analyzed in the data collection and processing module to extract the network features representing network status. Then, these features are analyzed in the anomaly detection model, which is built according to ML algorithms to classify the network status into anomaly or normality.

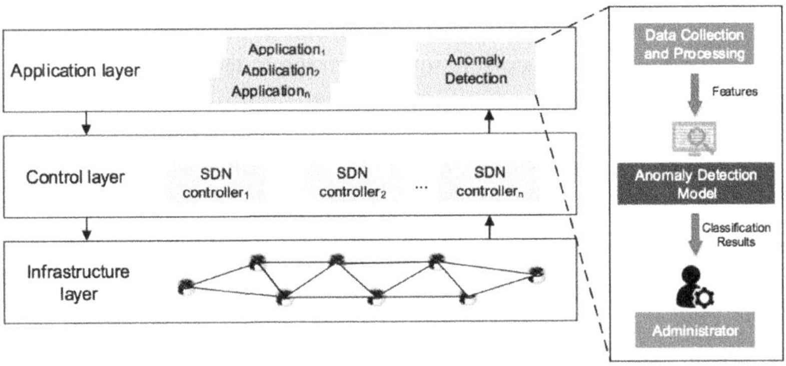

Figure 4.1. *Overall architecture of ML-based anomaly detection mechanism in the SDN environment. For a color version of this figure, see www.iste.co.uk/tong/troubleshooting.zip*

4.3.1. *ML-based anomaly detection method*

After collecting and extracting network features from the data collection and processing module, we analyze these features using the ML-based anomaly detection method to identify the anomalies in the network. Network problems can lead to performance degradation and a fluctuation of network features. ML algorithms aim to learn this fluctuation to detect anomalies in the network. The detail of this approach is described in Figure 4.2.

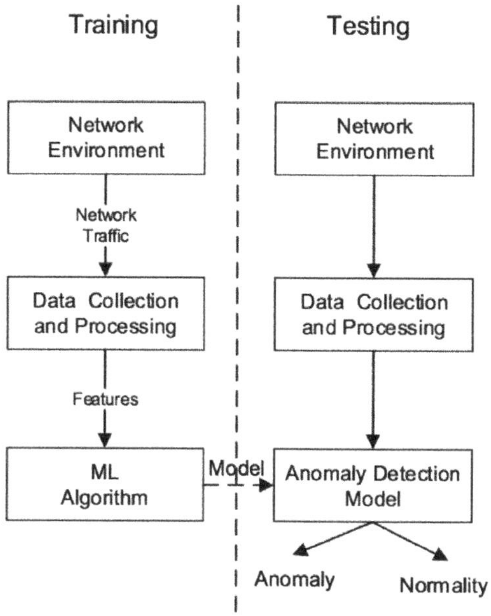

Figure 4.2. *The ML-based anomaly detection method.*

The two main phases in this method are training and testing. In the training phase, the network features will be analyzed according to an ML algorithm to obtain a pre-trained anomaly detection model for the testing phase. The objective of the anomaly detection model is to detect the anomalies in the network. There are different kinds of ML algorithms for anomaly detection. Consequently, we evaluate several algorithms to select the appropriate one and balance between accuracy and time complexity. The considered ML algorithms include multilayer perceptron (MLP) (Taud and Mas 2018), SVM (Ahmad et al. 2018), RF (Ahmad et al. 2018), Adaptive Boosting (AdaBoost) (Rehman Javed et al. 2020) and convolutional neural network (CNN) (Kwon et al. 2018). Concerning the CNN model, we use two convolutional 1D (one dimensional) layers, a max Pooling 1D, a flatter layer and three fully connected layers. For the others algorithms, we adjust hyperparameters such as depth of tree, number of estimators, etc. These configurations have been chosen based on existing studies (Hong et al. 2020).

4.3.2. *Data collection and processing*

This module collects and extracts network features representing network status (anomaly and normality) in the network. This module refers to the parameter measurement module in the troubleshooting framework (Figure 2.2). In this book, we consider congestion and simulate three root causes (link failure, switch failure and buffer overload) to obtain anomalous network status. We simulate link failure by generating a delay in links. Switch failure is simulated by generating rule failure, leading to a packet loss in the switches. Buffer overload is simulated by generating a massive amount of network traffic exceeding the link's capacity. The anomalous network statuses are simulated in two scenarios, in static and dynamic networks. In the static network, delay and loss parameters in the links are tied to a particular value. In contrast, in the dynamic network, link status is changed according to the Gilbert–Elliot (GE) model (Bildea et al. 2015) following a probability value after a given time (400 s). The probability of changing from normal to anomalous status and anomalous to normal status are p and r, respectively. If link status is anomalous, delay and loss are changed as in Table 4.1.

Status		**Static network**	**Dynamic network**
Anomalous status	Link failure	Delay: 125 ms	Delay: [25, 50, 75, 100, 125 ms]
	Switch failure	Loss: 50%	Loss: [10, 20, 30, 40, 50%]
	Buffer overload	Bandwidth: 10 mbps, Sending rate: 60–100 mbps	Bandwidth: 10 mbps, Sending rate: 60–100 mbps
Normal status		Sending rate: 1–3 mbps	

Table 4.1. *Considered network conditions*

In each link, we extract three network parameters including latency, packet loss and link utilization. These parameters are discussed in the parameter measurement module (see section 2.3.3). This choice is explained by the fact that these parameters are taken into account in many service-level agreements (SLA) of service providers (Frost and Bryant 2011). The parameters on each link will be aggregated to calculate the parameter of a routing path (network features), which is used for anomaly detection. After that, these features are normalized into a value between 0 and 1 according to max normalization

(Singh and Singh 2020). In anomaly detection, we consider time-series features, and a sample aggregates features of 10 consecutive time steps to identify the anomaly. An input sample is a matrix 10×3 (number of time steps × number of features). The anomaly detection datasets are depicted in Table 4.2.

Status	Static Network	Dynamic Network
Anomalous	24452	44172
Normal	11075	11075

Table 4.2. *Anomaly detection datasets*

4.4. Experimental results

4.4.1. *Experimental setup*

In the experiments, *mininet* (Dholakiya et al. 2021) is used to generate network topology with five spine nodes and two leaf nodes. The delay and loss parameters on each link are configured in *mininet* based on the *TCLink* argument. In the GE model, p and r are set to 0.81 and 0.07 as in Bildea et al. (2015). The ML algorithms are implemented according to the *scikit-learn* library (Pedregosa et al. 2011) in Python. The training, validation and testing phase comprise 76%, 4% and 20% of the anomaly detection dataset, respectively.

The performance of the ML algorithms is evaluated according to performance metrics: precision, recall and F1-score (F-measure) (Sokolova and Lapalme 2009). Precision is the percentage of relevant flows retrieved, while recall is the percentage of retrieved flows that are relevant. The F1-score (F-measure) represents a harmonic mean between precision and recall. The details of these parameters are depicted in equations [4.1]–[4.3]. There are two ways to evaluate the quality of the overall classification: the micro-averaging and the macro-averaging value. In macro-averaging, a metric is averaged over all classes that are treated equally, whereas micro-averaging is based on the cumulative True Positive (TP), False Positive (FP), True

Negative (TN) and False Negative (FN) of the dataset.

$$Precision = \frac{TP}{TP + FP}. \quad [4.1]$$

$$Recall = \frac{TP}{TP + FN}. \quad [4.2]$$

$$F1 - score = \frac{2}{1/Recall + 1/Precision}. \quad [4.3]$$

4.4.2. *Performance analysis*

There are two anomaly detection datasets in static and dynamic networks. We consider a balance between the performance and time complexity of ML algorithms to select an appropriate one. Table 4.3 shows the performance metrics of ML algorithms for the dataset in a static network. In the dataset in a static network, the range of network features between classes sometimes overlaps, so this leads to noise in the dataset. The performance metrics of SVM are the lowest in four algorithms. The micro- and macro-precision, recall and F1-score are approximately 83.3%. The reason for this is that SVM does not perform well when the dataset contains more noise (e.g. classes overlap, etc.) (Gales et al. 2009). RF builds multiple decision trees on different subset of the dataset and selects the class selected by most trees. The number of decision trees in RF is set to 100 as default in the *scikit-learn* library. Consequently, RF can achieve good classification results. The precision, recall and F1-score of RF algorithm are close to 98.8%. AdaBoost is an ML algorithm using the boosting ensemble method and a decision tree as the individual model. The boosting ensemble method learns the individual model from the mistakes of previous models. Despite learning from the mistakes by increasing the weight of misclassified samples, it can lead to uncertain results when a dataset contains noise. Therefore, the precision, recall and F1-score of AdaBoost reduce from 1% to 1.5% in comparison with the RF algorithm. The performance of CNN is the highest in four algorithms, with approximately 99%.

Table 4.4 indicates the running time of the ML algorithm for the dataset in a static network. These values are the average training and testing times for a sample in the training and testing dataset. SVM has the largest training and testing times, at 0.9183 and 0.3940 ms, resepectively. The testing time of RF and AdaBoost are nearly equal to 0.009 ms. Although CNN has the highest performance, its training and testing times are higher than the figures for RF and AdaBoost. In a fold, the training time of an epoch of CNN is 0.0548 ms. This value increases against the number of folds and epochs. In the book, we configure the number of folds and epochs to 20. Therefore, the total training time for a sample is 21.92 ms (0.0548·20·20). The testing time of CNN nearly doubles the figures for RF and AdaBoost.

Class	**Precision**				**Nb samples**
	SVM	RF	Adaboost	CNN	
Normal	0.7240	0.9881	0.9587	0.9830	2215
Anomalous	0.8950	0.9880	0.9834	0.9965	4891
Micro	0.8379	0.9880	0.9757	0.9923	7106
Macro	0.8095	0.9881	0.9710	0.9898	7106

Class	**Recall**				**Nb samples**
	SVM	RF	Adaboost	CNN	
Normal	0.7756	0.9734	0.9634	0.9923	2215
Anomalous	0.8661	0.9947	0.9812	0.9922	4891
Micro	0.8379	0.9880	0.9757	0.9923	7106
Macro	0.8209	0.9840	0.9723	0.9923	7106

Class	**F1-score**				**Nb samples**
	SVM	RF	Adaboost	CNN	
Normal	0.7489	0.9807	0.9610	0.9876	2215
Anomalous	0.8803	0.9913	0.9823	0.9944	4891
Micro	0.8379	0.9880	0.9757	0.9923	7106
Macro	0.8146	0.9860	0.9717	0.9910	7106

Table 4.3. *Performance metrics of ML algorithms in anomaly detection for the dataset in a static network*

Algorithms	**Training Time (ms)**	**Testing Time (ms)**
SVM	0.9183	0.3940
RF	0.2040	0.0098
AdaBoost	0.1548	0.0094
CNN	0.0548	0.0163

Table 4.4. *Time complexity of ML algorithms in anomaly detection for the dataset in a static network*

Class	Precision				Nb samples
	SVM	RF	Adaboost	CNN	
Normal	0	0.9738	0.9060	0.9774	2215
Anomalous	0.7995	0.9850	0.9773	0.9946	8834
Micro	0.7995	0.9828	0.9630	0.9911	11049
Macro	0.3998	0.9794	0.9417	0.9860	11049
Class	**Recall**				**Nb samples**
	SVM	RF	Adaboost	CNN	
Normal	0	0.9395	0.9097	0.9783	2215
Anomalous	1	0.9937	0.9763	0.9943	8834
Micro	0.7995	0.9828	0.9630	0.9911	11049
Macro	0.5000	0.9666	0.9430	0.9863	11049
Class	**F1-score**				**Nb samples**
	SVM	RF	Adaboost	CNN	
Normal	0	0.9563	0.9079	0.9779	2215
Anomalous	0.8886	0.9893	0.9768	0.9945	8834
Micro	0.7995	0.9828	0.9630	0.9911	11049
Macro	0.4443	0.9728	0.9424	0.9862	11049

Table 4.5. *Performance metrics of ML algorithms in anomaly detection for the dataset in a dynamic network*

Algorithms	Training Time (ms)	Testing Time (ms)
SVM	0.8439	0.4344
RF	0.1810	0.0103
AdaBoost	0.1615	0.0092
CNN	0.0403	0.0165

Table 4.6. *Time complexity of ML algorithms in anomaly detection for the dataset in a dynamic network*

Tables 4.5 and 4.6 illustrate the performance metrics and time complexity of ML algorithms for the dataset in a dynamic network. There is an imbalance between the number of normal and anomalous network status, so the normal network status is classified as an anomalous network status in the SVM algorithm. Therefore, the precision, recall and F1-score of normal network status are 0 in the SVM algorithm. Moreover, CNN has the highest performance metrics in the four algorithms, but its time complexity nearly doubles the figure for RF and AdaBoost. Regarding the balance between the performance metrics and the running time of ML algorithms, we consider RF as an ML algorithm in the ML-based anomaly detection method.

4.5. Conclusion

This chapter presents an anomaly detection approach using an ML algorithm to detect the anomalies in the network. Our contribution consists of an anomaly detection approach containing two main modules: (i) data collection and processing and (ii) anomaly detection model. The first module collects and extracts network features representing network status. Then, these features are analyzed in the anomaly detection model, which is learned according to an ML algorithm. There are many ML algorithms, so we evaluate the performance and time complexity of these algorithms to select an appropriate ML algorithm. As a result, the RF algorithm is selected as a ML algorithm in anomaly detection. The experimental results show that the micro and macro F1-score can achieve up to approximately 99% in the considered datasets.

After identifying the anomalies in the network, we need to implement a basic remediation approach to reduce the negative impacts of anomalies and guarantee the availability of the network until the root cause of the anomalies is identified and solved. Therefore, in Chapter 5, we present a QoE-based application-aware segment routing to make appropriate routing strategies corresponding to different applications and meet strict user requirements for encrypted traffic.

5

Temporary Remediation: SDN-based Application-aware Segment Routing for Large-scale Networks

"The Internet is not just one thing"

James H. Clark

Chapter 5 presents a QoE-based application-aware segment routing in the SDN environment for encrypted traffic. This approach is implemented in the temporary remediation module, which is briefly depicted as in Figure 2.2.

5.1. Introduction

According to related work (Fonseca and Mota 2017; Yu et al. 2018), there are many remediation approaches including load balancing, routing, etc. Moreover, many studies have proven the performance of routing in problem remediation (Kim et al. 2016; Jurkiewicz et al. 2020). Therefore, we consider an application-aware routing in the temporary remediation module. The application-aware routing, which implements different kinds of routing policies corresponding to differentiated applications, can overcome this drawback. To facilitate the application-aware routing, two requirements related to a routing technique supporting traffic engineering and a network architecture that can obtain a global view of network need to be met.

For the first requirement, multi-protocol label switching (MPLS) is a potential solution deployed by network operators to improve their IP networks. However, MPLS contains three main drawbacks (Ventre et al. 2020; Abdullah et al. 2018). Firstly, it requests an IP network to maintain an explicit state at network nodes along an MPLS path, bringing a scalability problem in both the control and data plane. Secondly, MPLS cannot benefit from the load balancing given by equal-cost multi-path routing (ECMP). Finally, MPLS with the support of Interior Gateway Protocols (IGPs) for routing protocol cannot be easily implemented on multiple autonomous systems (ASs). Therefore, many network operators (e.g. NTT, Vodafone, etc.) implement segment routing (SR) in their network infrastructure as a solution for these issues (Cisco 2018, 2020). The core idea of SR architecture is based on the notion of source routing (Kushwaha et al. 2020) and tunneling to guarantee the scalability property in decreasing the amount of state information to be processed in the core network. Besides, SR's main benefit is to fix the scalability issues and limitations of the MPLS approach. Concretely, SR does not require any state maintenance in core network nodes. Moreover, it takes advantage of the ECMP routing and the implementation on multiple ASs (Kukreja et al. 2016).

For the second requirement of the application-aware routing, software-defined networking (SDN) is a promising solution. It decouples the control layer from the infrastructure layer, offering an opportunity to obtain a global network view. Therefore, application-aware SDN-based SR is concerned by the research community (Kukreja et al. 2016; Peng et al. 2020). However, this SR mechanism contains two main disadvantages:

– Firstly, this routing mechanism considers the application class to meet different service-level agreement (SLA) requirements, but application identification is sometimes complicated due to encrypted traffic. In the past, the application class can be obtained by Differentiated Services Code Point (DSCP) in packet's header or deep packet inspection. However, DSCP field can be changed during packet's transmission (Custura et al. 2018) while the deep packet inspection is not effective with encrypted traffic (Dong et al. 2020). Indeed, many service providers encrypt their data during the transmission to protect the user's privacy. According to a recent Cisco report (Cisco 2021a), 80% of web traffic was encrypted by 2019 in comparison to 40% by 2016. Consequently, there is a necessity for a novel traffic classification approach to identify the application class in this case. Much existing research work focus on the classification approaches for VPN (Wang et al. 2017) and TLS/SSL (Wang et al. 2018) traffic. A novel solution is studied here to obtain the application class for the traffic of Quick UDP Internet Connection (QUIC) (Langley et al. 2017), which is a new transport layer network protocol developed by Google from 2012. The amount of QUIC traffic comprises 35% of Google's egress traffic (approximately 7% of global Internet traffic) and will continue to increase in the future (Lustig 2021).

– Secondly, routing paths are identified according to the human intervention, so it is not adaptive with the unprecedented change of network environments. For the past few decades, quality of service (QoS) played an essential role in the network systems, so many studies, such as Lee and Sheu (2016), concentrated on QoS-aware SR mechanisms. The main objective is to optimize the network resource utilization as well as meet network requirements related to QoS. Nevertheless, selecting appropriate QoS parameters is sometimes complicated. Therefore, quality of experience (QoE), which network operators need to assure, is considered in SR in this work. QoE is a metric correlated to QoS metrics, but it is a perception of end-user. Consequently, it facilitates network policies to guarantee the user's SLA requirements.

In this chapter, a novel SDN-based SR mechanism for encrypted traffic is presented for network operators. The objective is to implement the corresponding routing policies for various SLA requirements because each application has a specific SLA requirement. This mechanism selects the appropriate paths in a particular routing policy using reinforcement learning (RL) and the network environment's feedback (QoE) to adapt to dynamic network conditions. There are traditional algorithms in the QoE-aware SR scheme (e.g. selecting a path with maximal QoE value, etc.). However, it is not effective with dynamic networks. With the development of 5G, the networks have become more and more complex. Using the RL algorithm in the QoE-aware SR mechanism offers opportunities to select appropriate paths to adapt to the dynamic changes of network environments and improve long-term performance.

Outline: the remainder of the chapter is structured as follows. Section 5.2 introduces the related work in SR. In section 5.3, the chapter presents the novel SR mechanism. Section 5.4 describes the experimental results of this mechanism. Finally, the chapter concludes with section 5.5, which highlights our future work.

5.2. Application-aware routing mechanisms

This section presents related work on application-aware routing, application-aware MPLS, application-aware SR mechanisms and highlights its pros and cons.

5.2.1. *Application-aware routing*

Li et al. (2020) emphasized that network operators have encountered a challenge of providing better services with the rapid growth of 5G and multimedia services, which requires diverse network requirements (e.g. low latency, high reliability, etc.). However, network operators are unaware of which applications in their networks, so they cannot obtain a global view to manage the networks effectively. Therefore, many studies focus on application-aware routing techniques in network management. Application-aware routing is a routing technique that selects optimal routes corresponding to an individual application thanks to a corresponding routing strategy (Pongsakorn et al. 2017).

Adami et al. (2015) proposed an application-aware routing technique for Session Initiation Protocol (SIP) traffic. Firstly, this approach uses a deep packet inspection to identify SIP flows using its signatures (e.g. UDP packet, port 5060, content type: application/sdp, etc.). Then, this approach selects the shortest paths for these SIP flows using the Dijkstra algorithm with link utilization as a constraint.

Cheng et al. (2015) proposed an application-aware routing algorithm to deploy various routing policies corresponding to different applications. This algorithm classifies network traffic into three categories: real-time applications (e.g. VoIP, gaming, etc.), streaming applications (e.g. video streaming, web browsing, etc.) and miscellaneous applications (e.g. file sharing, etc.). Then, each category is processed with a specific routing policy using different network parameters such as link load, delay and delay variation.

Pongsakorn et al. (2017) proposed Overseer, an application-aware routing in SDN. Overseer analyzes network traffic using a deep packet inspection to classify into three kinds of applications including bandwidth-oriented (e.g. HTTP, FTP, media streaming, etc.), latency-oriented (e.g. SSH, online game, etc.) and default applications. Then, Overseer implements different routing policies corresponding to different kinds of applications containing maximal-bandwidth, minimal-latency and minimal-path-length strategies. The first one is to forward packets of bandwidth-oriented applications through the paths with the largest bandwidth while the second one aims to forward packets of latency-oriented applications through the paths with the smallest latency. The final one is to implement a default strategy that selects the shortest paths to forward packets of default applications.

Jeong et al. (2017) proposed an application-aware traffic engineering system in SDN to implement differentiated treatments on various applications. This approach uses a deep packet inspection to identify application classes and sends application information to a Traffic Scheduler. In this module, each application is assigned to a corresponding priority queue. According to application-specific priority queues and current network states, Traffic Scheduler calculates optimal routes corresponding to different applications. In this approach, the authors take into account two kinds of traffic including iPerf and YouTube.

Rego et al. (2019) proposed an improvement of Open Shortest Path First (OSPF) routing protocol in SDN environments. This approach changes the metric calculation (e.g. bandwidth, delay, etc.), adapting to different applications to select appropriate routing paths. However, identifying a practical metric calculation is sometimes complicated.

The above approaches identify the application class using the deep packet inspection, but it is not effective with encrypted traffic. Moreover, these approaches use an IP routing mechanism that identifies next-hops based on a destination IP address. It requires much processing time because each switch needs to analyze the packet's header to obtain the IP address. Besides, there is a large number of rules to be installed in the switches due to the rapid growth of Internet traffic, so these mechanisms require more ternary content-addressable memory (TCAM) resource consumption (Barakabitze et al. 2018).

5.2.2. *Application-aware MPLS*

The emergence of MPLS is a promising solution to overcome the drawbacks of IP routing mechanisms. MPLS identifies the next-hops based on the label added to the packet, so this requires less processing time and TCAM resource consumption.

Bahnasse et al. (2018) proposed an application-aware MPLS in SDN to optimize network resources. This approach identifies VoIP, video, HTTP and ICMP traffic based on DSCP field in the packet's header. Then, a specific routing policy is applied for each application to meet bandwidth constraints. However, identifying the application class using DSCP sometimes is inaccurate because it can be changed during data transmission (Custura et al. 2018).

Google (Yap et al. 2017) proposed Espresso, a SDN-based Internet peering edge routing infrastructure that offers an opportunity for the application-aware MPLS mechanism at the Internet-peering scale. According to integrating application-aware MPLS mechanism, Espresso delivers 13% more network traffic on their infrastructures and improves link utilization and user perception compared to Border Gateway Protocol (BGP)-based routing. Nevertheless, MPLS suffers from several hindrances related to scalability problems and ECMP routing.

5.2.3. *Application-aware SR*

Many other proposals on SR are proposed to address the hindrances of MPLS. Kukreja et al. (2016) presented a demonstration of SDN-based SR for multi-domain networks. In this demonstration, an orchestrator finds suitable routing paths and encodes to packet's header corresponding to diverse applications. The objective is to meet different resource requirements of these applications (e.g. bandwidth, delay constraints, etc.). Nevertheless, it is assumed that the orchestrator knows the class application.

Peng et al. (2020) proposed an application-aware network framework that takes advantage of SR to meet their SLA requirements. This framework identifies the application characteristics according to the deep packet inspection mechanism and then forwards packets into corresponding paths (policy or traffic engineering tunnel). Nevertheless, these SR mechanisms identify the routing paths because of human intervention, which is ineffective with the unprecedented change of network environments. In contrast, the proposed SR mechanism identifies the routing paths using RL to adapt to dynamic networks. Moreover, classifying the network traffic using deep packet inspection is not practical due to encrypted traffic nowadays (Dong et al. 2020). Therefore, the proposed SR mechanism implements a novel traffic classification approach to classify the encrypted network traffic and identify the application classes.

5.3. Adaptive segment routing mechanism for encrypted traffic

5.3.1. *Overview of the SDN-based adaptive segment routing framework*

The multi-modular system is depicted in Figure 5.1. During the transmission from servers to clients, a network flow is transmitted from ingress switches to an sFlow collector (Ujjan et al. 2020) to obtain the class of application because of traffic classification module. In the beginning, the network flow is forwarded using the shortest paths from the standard SR algorithm. After that, network parameters from parameter measurement module are analyzed in anomaly detection module to detect abnormal symptom of problems (e.g. increase of latency, packet loss, etc.). When the problems occur in the network, the network flow is forwarded using an adaptive SR mechanism that selects the appropriate routing paths to meet strict user requirements related to the QoE corresponding to each network application. When the problems are solved, the network flow will be forwarded using the standard SR algorithm. The major components in this framework are described as follows.

– *Network monitoring* includes two essential modules containing traffic classification and parameter measurement modules. The former aims to identify the class of application on network flows (video streaming, file transfer and VoIP) in the context of encrypted traffic while the latter aims to monitor and collect network parameters for further modules.

– *Anomaly detection* takes into account time-series network parameters to detect an abnormal symptom of network problems.

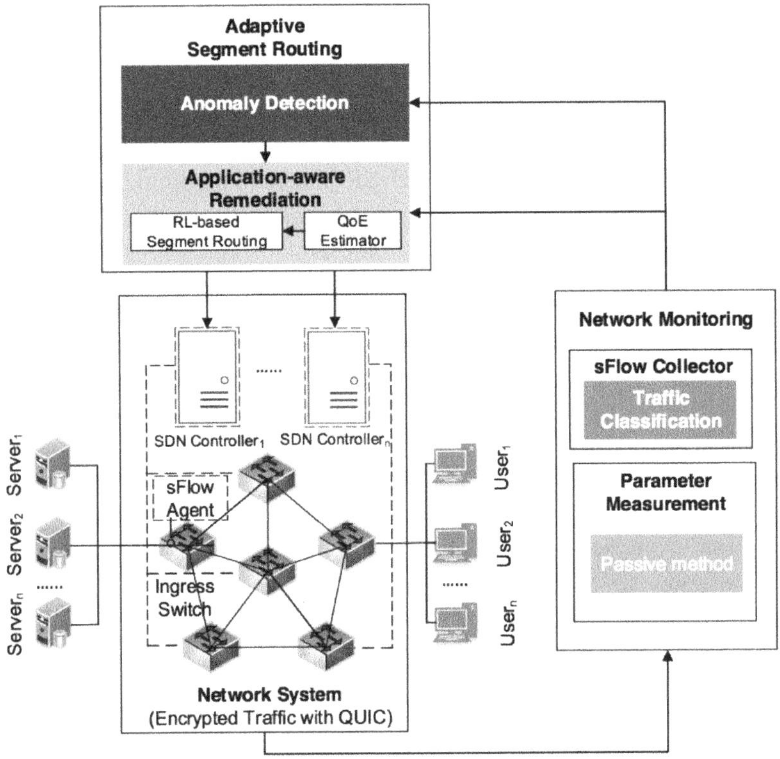

Figure 5.1. *The SDN-based adaptive SR framework issued from the global troubleshooting framework. For a color version of this figure, see www.iste.co.uk/tong/troubleshooting.zip*

– *Application-aware remediation* is used to consider various routing strategies corresponding to different kinds of applications to optimize the QoE in the network. This module is considered as a temporary remediation module to decrease negative impacts of anomalies (e.g. high loss, high latency, etc.) until its root causes are solved completely.

The new SR mechanism considers the class of application to implement appropriate routing policies corresponding to each kind of application. When network traffic is encrypted, this information is hidden. Therefore, there is a necessity of a traffic classification module for identifying this information. The traffic classification is presented in the following section.

5.3.2. *Network monitoring*

5.3.2.1. *Traffic classification*

A novel traffic classification approach is presented in this section to identify different kinds of applications for encrypted traffic. In Cisco (2021b), it is reported that video streaming, file sharing and VoIP will have comprised over 80% of global IP traffic by 2022. Consequently, these applications are considered in the traffic classification module. Regarding the network traffic for traffic classification, it can be collected directly from OpenFlow switches, but it is not effective with a huge amount of network traffic. Therefore, sFlow (Ujjan et al. 2020), a standard for network monitoring supporting packet sampling technique, is deployed to reduce the collected traffic volume and offers opportunities to implement the traffic classification module in real time. When the network traffic traverses the ingress switches (edge switches), the sFlow agents send the network traffic to the sFlow collector after sampling. At the sFlow collector, the traffic classification module collects the network traffic to identify the class of application. This module is described as follows (Figure 5.2).

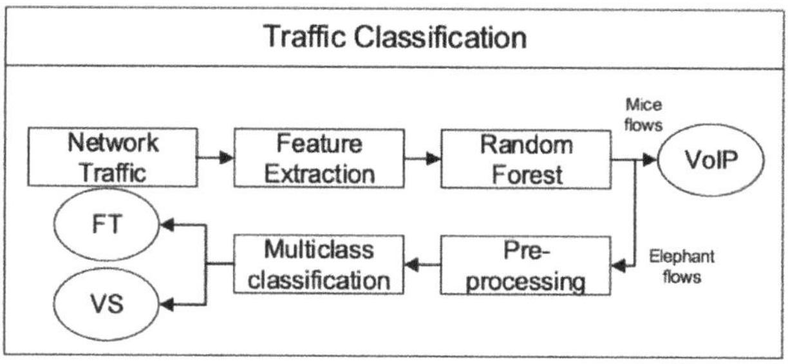

Figure 5.2. *The novel traffic classification approach for encrypted traffic*

There are two kinds of flows in the network traffic including mice flows (small continuous flows in total bytes) and elephant flows (huge continuous flows in total bytes). After investigating their characteristics, flow-based features (handcrafted features) of the first few packets in each flow (Lopez-Martin et al. 2017) are collected and analyzed using the random forest algorithm (Zhou et al. 2020) to classify the network traffic into the mice flows (VoIP) or the elephant-flows (video streaming and file transfer). This approach considers the random forest algorithm in mice-flow identification over several traditional machine learning algorithms (e.g. SVM, MLP, etc.) based on the research in Zhou et al. (2020). After that, the elephant flows are classified into video streaming (VS) or file transfer (FT) using packet-based features (implicit features) and the convolutional neural network (CNN). The latter learning algorithm is taken into account in this approach because it contains characteristics including spare connectivity, parameter sharing and equivariant representations. This helps to learn more effective representations in comparison with traditional machine learning algorithms (Goodfellow et al. 2016). The detail of the traffic classification module is described in our previous work (Tong et al. 2018; Chapter 3).

Regarding time complexity of this module, it is based on CNN, and the total time complexity of all convolutional layers (He and Sun 2015) is estimated by equation [5.1]:

$$O(\sum_{l=1}^{d} n_{l-1} \times s_l^2 \times n_l \times m_l^2). \qquad [5.1]$$

where d is the number of convolutional layers, n_{l-1} is the number of input channels of the lth layer, n_l is the number of filters in the lth layer, s_l is the spatial size of the filter and m_l is the spatial size of the output feature map.

This time complexity is applied in both training and testing phase, and training time is approximately three times testing time.

5.3.2.2. *Parameter measurement*

Many SLAs of service providers depend on several performance metrics such as latency, packet loss and link utilization (Abdullah et al. 2018; Bahnasse et al. 2018). Consequently, a parameter measurement module is implemented to measure these parameters on each link in the network. Latency is measured according to an existing work (Li et al. 2018), while packet loss and link utilization are measured because of *PortStatistics* API Team (2021) in the controller.

5.3.3. *Anomaly detection*

In this section, an anomaly detection approach is presented to identify abnormal symptoms of network problems. Unlike the existing rule-based approaches that trigger the alarms when network parameters (e.g. packet loss, delay, etc.) exceed a threshold, its fluctuations are monitored to early detect the unusual symptoms in this approach (Lee et al. 2017; Maimó et al. 2018). The time-series parameters on network flows such as latency, packet loss and link utilization are taken into account. These parameters are concatenated into a one-dimensional vector. Then this vector is analyzed according to the random forest algorithm as in existing work (Wang et al. 2020; Hong et al. 2020) to classify the network states into normal or abnormal states to early detect the anomaly of network problems.

The time complexity of random forest algorithms in training phase is $O(n_0 \times log(n_0) \times d \times k)$, where n_0 is the number of input sample, d is dimensional of data and k is the number of trees in the forest. Besides, the time complexity in testing phase is $O(k \times d')$, where d' is the depth of tree (Gao et al. 2018).

5.3.4. *Application-aware remediation*

5.3.4.1. *QoE estimator*

The rapid growth of the Internet leads to the diversity of multimedia applications. Consequently, deploying a general QoE model for different applications is ineffective due to its various QoS requirements. Therefore, we implement a novel QoE estimator (Figure 5.3) to calculate QoE of different applications. The application class is identified in the traffic classification module. Then, a specific QoE model is used to calculate QoE for a particular application.

QoE can be calculated according to subjective and objective approaches. The former requires participants to evaluate their mean opinion score (MOS) about multimedia applications, while the latter builds an objective model calculating MOS using parameters (e.g. network parameter, application parameter, etc.). However, the subjective approach is costly and ineffective with real-time monitoring, whereas identifying an effective model in the objective approach is sometimes complicated. Therefore, we implement a hybrid approach that uses a subjective dataset to learn QoE models based on ML algorithms. The hybrid approach offers opportunities to estimate MOS in real time. We use an MOS dataset, which is built in a collaboration between our laboratory (LISSI) and Orange (a network operator in France) (Amour et al. 2015a). There are many ML algorithms (e.g. MLP, SVM, random forest, etc.). The random forest has less root mean square error in comparison with others (Amour et al. 2019), so it is considered as an ML algorithm in QoE estimation. MOS is calculated because of network parameters (e.g. latency, packet loss, etc.). There are five levels of MOS including 1 (bad), 2 (poor), 3 (fair), 4 (good) and 5 (excellent). The detail of this module is described in (Amour et al. 2019).

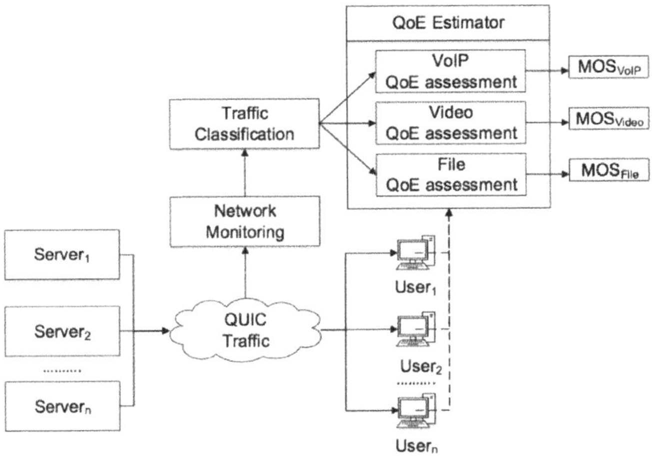

Figure 5.3. *The QoE estimator for encrypted traffic*

5.3.4.2. *RL-based segment routing*

The SR algorithm is formalized as an RL task that contains *agent, state, action, reward* and *policy* (Figure 5.4). The details are described as follows.

– *Agent*: an entity in the network system applies a learning algorithm to perform its tasks. In a routing problem, an agent selects appropriate paths to optimize a reward.

– *State s*: a snapshot of the network environment which is observed by the agent.

– *Action a*: an action illustrates how an agent replies to the network environment. In the routing problem, the action is a routing path between a server and a client in the network. All routing paths in the network can be obtained in the SDN controller.

– *Policy* π: a policy is a map from an observed state to action in the network environment.

– *Reward* r: a reward is a feedback of the network environment corresponding to the agent. In the routing problem, the agent monitors a network state s and performs an action a from the routing policy. Then, the agent moves to next state s' and receives a reward r. The reward is the MOS score of a chosen path, which is calculated via the pre-trained QoE estimator (section 5.3.4.1).

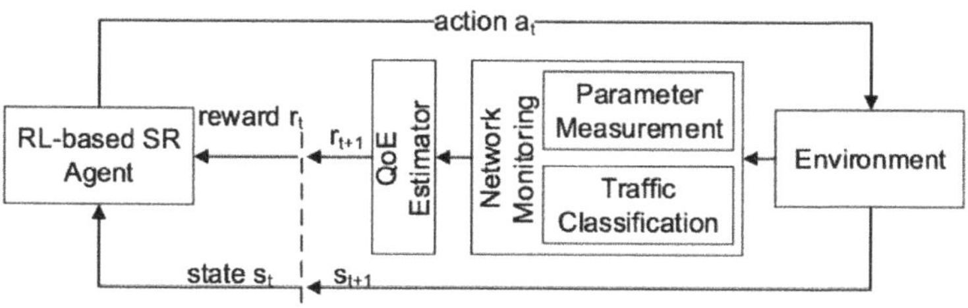

Figure 5.4. *The RL-based SR mechanism*

The main goal of RL is to optimize an objective function O_f (equation [6.4]):

$$O_f = Max\ E[\sum_{t=0}^{\infty} \gamma_t \times r_t]. \qquad [5.2]$$

where $\gamma_t \in [0, 1]$ is a discount factor.

In RL, there are two kinds of approaches including model-based and model-free approaches. In the first approach, the agents learn the environment model and enhance its policies to obtain optimality while the agents optimize its policies without prior information about the network environment in the second approach. The first one can learn faster than the second one. It is still less popular because of a large storage cost and dependence on accuracy of an initial information (Mammeri 2019). Therefore, the model-free approach is used in this book. In this approach, Q-value estimates how good it is to execute a given action in a given state. $Q(s, a)$ is the expected return starting from state s and taking action a following policy π. At a time step, the agent is in state s, performs action a, receives reward r and moves to next sate s'. The Q-value is updated as in equation [6.5]:

$$Q(s,a) = (1-\alpha)Q(s,a) + \alpha[r + \gamma \max_{a'} Q(s',a')]. \quad [5.3]$$

where α is a learning rate and γ is a discount factor.

Time complexity of reinforcement learning algorithm is $O(m \times N_0)$, where N_0 is the number of action and m is the size of state space (Koenig and Simmons 1993; Jin et al. 2020).

5.3.4.3. *Exploration–exploitation trade-off*

An exploration and exploitation phase in RL needs to be balanced to obtain an optimal cumulative MOS score (cumulative reward). The exploitation phase, which selects the routing path (action) with maximal Q-value, cannot be implemented systematically because each routing path needs to be evaluated frequently to achieve the optimal MOS score. In this chapter, the trade-off between the exploration and exploitation phase is formalized as a multi-armed bandit (MAB) problem. MAB problem is a formalization of sequential decision-making tasks. At a time step, a decision-maker selects an action and receives a reward from an unknown distribution corresponding to this action. The main objective is to maximize the total reward received through a sequence of actions. In this chapter, three selection algorithms are presented to resolve the MAB problem: *ϵ-greedy* (Temlyakov 2015), *softmax* (He et al. 2018) and *UCB1* (upper confidence bounds) (Tran et al. 2013).

Firstly, *ϵ-greedy* is the simplest algorithm to resolve the bandit problem. Concretely, the agent selects the routing path with the highest Q-value with a probability of (1-ϵ). Otherwise, the agent selects the routing path randomly. Then, the ϵ value reduce with the time so that the agent can learn more about the network environment and become more confident.

Secondly, the *softmax* algorithm selects the routing paths according to a probability function of Q-value. Each routing path a_i is assigned to a probability p_i as in equation [5.4]:

$$p_i = \frac{e^{\frac{Q_{a_i}}{\tau}}}{\sum_{j=1}^{N_0} e^{\frac{Q_{a_j}}{\tau}}} \quad [5.4]$$

where τ is a temperature parameter, N_0 is a number of routing paths and Q_{a_j} is a Q-value of routing path a_j.

When the temperature parameter τ is reduced, the routing paths are exploited more frequently. In that way, the temperature parameter τ is reduced to each episode (forwarding time). Therefore, *softmax* algorithm not only explores the less-used routing paths but also selects the best routing path in terms of expectation gains.

Finally, the *UCB1* algorithm is related to an index-based algorithm. UCB index is defined as a sum of a current Q-value and a confidence bound. The UCB index is described as in equation [5.5]:

$$UCB - index_{a_i} = Q_{a_i} + \sqrt{\frac{2ln(N)}{n_{a_i}}}. \qquad [5.5]$$

where Q_{a_i} is a Q-value of routing path a_i, n_{a_i} is a number of chosen time of routing path a_i and N is an episode number (forwarding time).

After calculating the UCB index for each routing path, *UCB1* algorithm selects the path with maximal UCB index. As shown in equation 5.5, the UCB index comprises two parts including Q-value Q_{a_i} and confidence bound $\sqrt{\frac{2ln(N)}{n_{a_i}}}$. A routing path is chosen when the Q value is large or the confidence bound is high. When the routing path with the large Q-value is chosen, this choice is an exploitation trial. When the confidence bound is high, this choice is an exploration trial. The confidence bound is higher when the number of chosen times of the routing path is smaller in comparison with other paths. In other words, the less routing path is selected, the more it has the opportunity to be selected.

5.4. Experimental results

5.4.1. *Experiment setup*

The performance of the new SR mechanism is evaluated via an emulation testbed with *mininet* v2.2 (Brian and José 2021) and *ONOS* (Open Network Operating System) controller v2.4. *mininet* is a popular emulator in the research community to emulate the network topology. ONOS only supports leaf-spine topology, so we use this topology from a simple scene (five spine nodes and two leaf nodes) to a complex one (more than 15 nodes). The number of nodes in the topology is set to 7, 17, 32 and 47. Moreover, *ONOS* also supports segment routing through *org.onosproject.segmentrouting* application, and we customize it to implement the novel SR mechanism. To generate QUIC traffic of video streaming in the network, we replay a pcap file of QUIC traffic from servers to clients using *tcpreplay* application, which is collected according to watching videos in YouTube with Google Chrome. The testbed is implemented in a PC described in Table 5.1. For the exploration–exploitation trade-off, ϵ and τ are set to 1 and 2, respectively. For the RL algorithm, α and γ are set to 0.7 and 1, respectively. These parameters are selected according to the existing work of Lin et al. (2016). The source code of the framework is available in Tong (2021c).

Operation system	**Ubuntu 16.04.6 LTS**
Processor	Intel(R) Core(TM) i5-6500 CPU @ 3.20GHz
Memory	2133MHz DDR4 8GB

Table 5.1. *Configuration of the PC used in the testbed*

The experiments are considered in four scenarios (Table 5.2). The network parameters are chosen to cover the maximum QoE range. The first scenario is to evaluate the performance of the proposal in the network condition without delay and loss, and other scenarios are used to consider this mechanism in the context with faults. Loss and delay parameters in the routing paths are generated according to *mininet*. In the last three scenarios, each routing path is randomly set to a specific delay, loss or both delay and loss. These delay and loss parameters are changed every 50 episodes, which are chosen according to the experiments to generate dynamic network states. An uniform link capacity is set to 10 Mbps, and a sending rate is set to 2.5 Mbps as in the existing work of Barakabitze et al. (2018).

The new application-aware SR mechanism is designed for three applications including video streaming, file transfer and VoIP. Among these applications, video streaming comprises the highest global IP traffic Cisco (2021b), so we consider video streaming as a proof of concept to thoroughly validate the performance of this SR mechanism.

<table>
<tr><th colspan="2">Scenarios</th><th>Descriptions</th></tr>
<tr><td colspan="2">perfect scenario</td><td>No delay and loss</td></tr>
<tr><td rowspan="3">Scenario with faults</td><td>delay</td><td>25, 50, 75, 100 and 125ms</td></tr>
<tr><td>loss</td><td>5, 10, 15, 20 and 25%</td></tr>
<tr><td>delay and loss</td><td>(25ms, 5% loss), (50ms, 10%), (75ms, 15%), (100ms, 20%) and (125ms, 25%)</td></tr>
</table>

Table 5.2. *Scenarios*

5.4.2. *Benchmark*

To validate the performance of the proposal, it is compared to the benchmarks, including:

– *Standard SDN-based SR (Standard_SR)*: this algorithm uses the Dijkstra algorithm to determine the shortest paths between servers and clients;

– *SDN-based SR with maximal QoE (Max_QoE)* (Grigoriou et al. 2017): this approach calculates the MOS score of all routing paths and selects the paths with maximal MOS score. In contrast, the MOS score of a chosen path is calculated in the proposal that helps to reduce resource consumption.

These approaches are evaluated via the following performance metrics:

– *MOS*: the perceived quality at the user's side;

– *CPU Usage*: the percentage of the CPU's capacity which is calculated via *ps* command (process status) in the Unix-like operation systems;

– *Control Overhead*: to discovery network topology and update status of links (e.g. latency, loss, etc.), a routing algorithm needs to generate the control packets (e.g. LLDP packets, etc.). Control overhead refers to the ratio of the control packet number to the total number of the sent packets.

5.4.3. *Performance analysis*

The selection algorithms for MAB formalization are presented including *UCB1*, *softmax* and ϵ-*greedy*. Their performances are first evaluated to select the appropriate selection algorithm. Then, the performance of our proposal is compared to the benchmarks including *Standard_SR* and *Max_QoE* mechanisms related to MOS score, CPU usage and control overhead. Besides, the running time of each module in the new SR mechanism is thoroughly evaluated.

5.4.3.1. *Time analysis*

In section 5.3, we analyzed the mathematical time complexity of algorithms in this SR mechanism. According to this time complexity, the traffic classification requires higher processing time in comparison with other modules. It requires 0.0316 ms to identify the network flows of VoIP and 3.4 ms to identify the network flows of video streaming and file transfer. To implement this module in real time, we use sFlow, a standard for network monitoring, to reduce the amount of network traffic collected for the traffic classification module when there is a huge amount of traffic in the networks. In the testbed, we configure sFlow agents to collect 10% of network traffic going through sFlow agents. Regarding the running time of the anomaly detection module, it requires 0.0127 ms to process a network flow. As for the RL-based segment routing module, it searches in the Q-table which represents routing policies, to select an appropriate path, so it can make instant decisions.

5.4.3.2. *Selection algorithms*

Figure 5.5 illustrates the average MOS score and standard deviation of *UCB1*, *softmax* and *ϵ-greedy* in the novel SR mechanism with four scenarios. In the *perfect scenario*, there is no delay and loss in the network. The sending rate is 2.5 Mbps while the link capacity is 10 Mbps. Consequently, changing the routing paths does not lead to the fluctuation of the MOS score in three selection algorithms. Although there is no significant difference between these algorithms, the MOS score of *softmax* is slightly better than the others. The average MOS score of *softmax*, *UCB1* and *ϵ-greedy* reach 4.47, 4.46 and 4.39, respectively. Besides, their standard deviations are 0.29, 0.291 and 0.31, respectively. In this scenario, the MOS score varies from 4.3 to 4.5. The MOS score of *ϵ-greedy* converges to 4.3, while the MOS score of the others converges to 4.5. Therefore, the average MOS score of *ϵ-greedy* is lower than two other selection algorithms.

The average MOS score of *UCB1* is the highest, and the standard deviation is the lowest in three selection algorithms for the other scenarios. The average MOS score of *UCB1* in *delay scenario*, *loss scenario* and the final scenario are 3.55, 3.7 and 3.5, respectively. *UCB1* first explores the routing path during the first few episodes and then converges to the optimal MOS score. When the network states are changed, *UCB1* can explore the less-used routing paths. Consequently, it continues to reach an optimal value quickly. *softmax* chooses the routing paths according to probability function, so it takes much time to converge to an optimal value. As a result, the MOS score of *softmax* is lower than the figure for *UCB1*. For ϵ-*greedy*, it first explores the routing paths with a high ϵ value. When ϵ value reduces to approximately 0, it cannot explore the routing paths frequently. Therefore, the MOS score of ϵ-*greedy* is lower than two other selection algorithms.

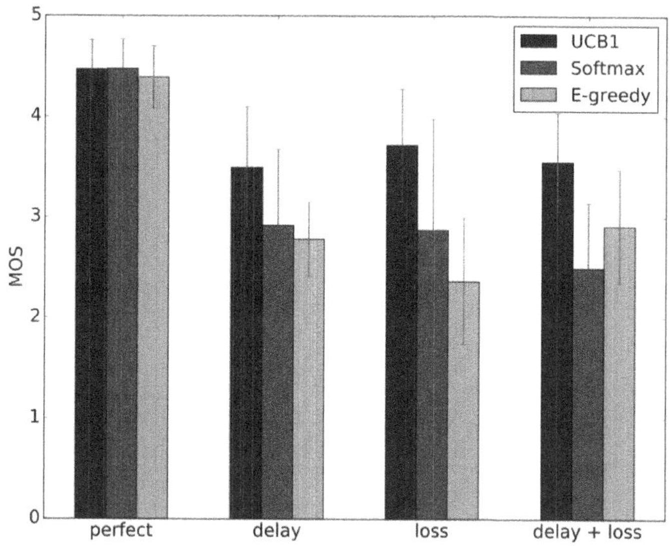

Figure 5.5. *Average MOS score and standard deviation of three selection algorithms in the novel SR mechanism. For a color version of this figure, see www.iste.co.uk/tong/troubleshooting.zip*

According to Figure 5.5, our proposal will implement different selection algorithms corresponding to various scenarios. *softmax* will be implemented for *perfect scenario*, while *UCB1* will be implemented for other scenarios in our proposal in the following experiments. Our proposal refers to *RL_Softmax_SR* in *perfect scenario* and *RL_UCB1_SR* for other scenarios.

5.4.3.3. *Segment routing mechanisms*

Figure 5.6 illustrates the MOS score against the episodes (forwarding time) of three SR mechanisms in four scenarios. In *perfect scenario* (Figure 5.6(a)), the role of the routing paths is the same due to no delay and loss in the network. As a result, the MOS score of three SR mechanisms is nearly equal. The average MOS score of our proposal (*RL_Softmax_SR*), *Max_QoE* and *Standard_SR* are 4.47, 4.47 and 4.46, respectively. Besides, their standard deviations are 0.29, 0.29 and 0.34, respectively.

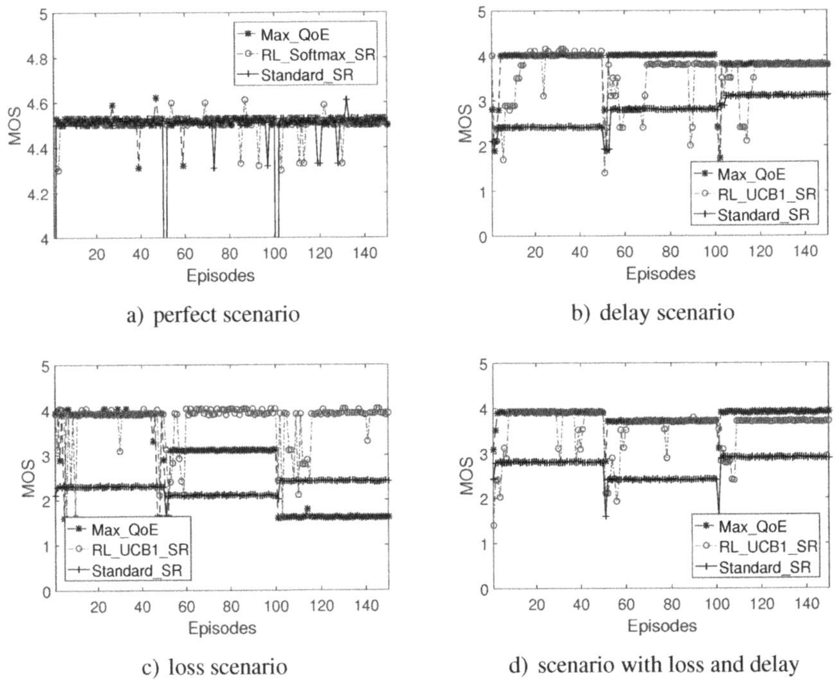

a) perfect scenario

b) delay scenario

c) loss scenario

d) scenario with loss and delay

Figure 5.6. *The MOS score of three SR mechanisms. For a color version of this figure, see www.iste.co.uk/tong/troubleshooting.zip*

In *delay scenario* (Figure 5.6(b)), each routing path is set to a specific delay parameter which leads to the differences in the MOS score between the routing paths. Therefore, there is a significant difference between three SR mechanisms. The MOS score of *Standard_SR* is the lowest in three SR mechanisms because it selects the shortest paths to forward the packets. The MOS score of this mechanism depends on the delay parameter, which is set to the shortest path. The network states are changed every 50 episodes, so the MOS score of this mechanism changes periodically. The MOS score of *Standard_SR* in the first, second and last 50 episodes are approximately 2.4, 2.8 and 3.1, respectively. *Max_QoE* monitors the network topology, calculates the MOS score of all routing paths and selects the path with the best MOS score. Therefore, it achieves a high MOS score (approximately 4) in the majority of episodes. In this scenario, our proposal (*RL_UCB1_SR*) implements *UCB1* to determine the routing paths to forward the packets. *UCB1* explores the routing paths in the first 16 episodes, so the MOS score varies between 1.7 and 4. After that, the MOS score of our proposal converges to an optimal value (approximately 4). At 50th and 100th episodes, the network states are changed to create dynamic network environments. UCB index of the less-used routing path becomes larger when a routing path is not chosen frequently. Consequently, *UCB1* can explore the less-used routing paths to select the appropriate path. As a result, the MOS score of our proposal can converge to optimal value quickly after a few episodes. Besides, our proposal can choose another routing path at the 24th and 89th episodes after converging to an optimal value. In other words, our proposal does not always select the routing path with the best UCB index to guarantee the exploration–exploitation trade-off. In Figure 5.6(b), the MOS score of our proposal is slightly lower or equal to the MOS score of *Max_QoE*.

In *loss scenario* (Figure 5.6(c)), the MOS score of *Standard_SR* is lower than the figure for our proposal. The MOS score of this mechanism in the first, second and last 50 episodes are 2.3, 2.1 and 2.4, respectively. Similar to previous scenarios, our proposal using UCB1 selection algorithm (*RL_UCB1_SR*) first explores the routing paths during the first few episodes and then converges quickly to optimal value (nearly 4). A remarkable feature from Figure 5.6(c) is that the MOS score of *Max_QoE* is lower than the MOS score of our proposal from the 50th episode. ONOS controller sends the control packets (e.g. LLDP packets, etc.) to discover the network topology every 3 s, so the delay on the routing paths is measured according to control packets as in the existing work of Liao and Leung (2016). Therefore, *Max_QoE* can monitor the network and select the path with the best MOS score in *delay scenario*. The loss parameter is measured via *PortStatistics* API in ONOS controller after data are transmitted in the network. Consequently, the loss on a routing path is not updated when this path is not chosen. At the beginning of the first 50 episodes in *loss scenario*, the loss of all routing paths initiated to 0. Therefore, *Max_QoE* can select the path with the best MOS score. At the 50th episode, the network states are changed. The routing paths with low MOS scores in previous network state becomes the paths with high MOS score in current network state, but the loss on these paths are not updated. Therefore, the MOS score of *Max_QoE* is lower than the figure for our proposal.

In *scenario with delay and loss* (Figure 5.6(d)), the MOS score of *Standard_SR* is the lowest in three SR mechanisms. Similar to *delay scenario*, the MOS score of our proposal (*RL_UCB1_SR*) is equal or slightly smaller than the MOS score of *Max_QoE*. The MOS score of *Max_QoE* in the first, second and last 50 episodes are 3.9, 3.7 and 3.9, respectively. Besides, the optimal MOS score of our proposal are 3.9, 3.7 and 3.7, respectively. Although the MOS score of *Max_QoE* is higher than the others, it requires much resource consumption related to CPU usage and overhead.

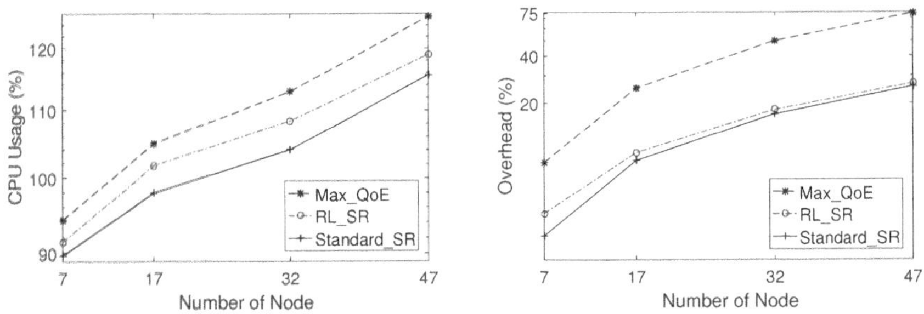

a) Average CPU usage

b) Average overhead

Figure 5.7. *The average CPU usage and overhead of three SR mechanisms. For a color version of this figure, see www.iste.co.uk/tong/troubleshooting.zip*

There is less difference between the CPU usage and overhead in four scenarios, so the average CPU usage and overhead of these mechanisms in four scenarios are depicted in Figure 5.7. The number of leaf nodes is increased while the number of spine nodes is not changed to obtain a larger network topology in these experiments. Figure 5.7(a) shows the CPU usage of three SR mechanisms. The testbed is implemented in a computer with four cores described in Table 5.1, so the CPU usage can reach a maximum of 400%. The CPU usage of *Max_QoE* is the highest in three SR mechanisms because it monitors the entire network topology in order to obtain the routing path with the best MOS score. The CPU usage of *Max_QoE* for 7 nodes is 94.3%, and it increases to 125.48% for 47 nodes. The CPU of our proposal is lower than the figure for *Max_QoE*. It raises from 91.45% to 118.95% when the number of nodes increases from 7 to 47. Besides, there is an increase in the CPU usage of *Standard_SR* from 89.73% to 115.63% when the number of nodes raises from 7 to 47.

Figure 5.7(b) indicates the control overhead of three SR mechanisms. The overhead of *Max_QoE* is higher than two other mechanisms because *Max_QoE* sends the control packets to monitor the entire network topology. It increases rapidly from 8.33% to 75% when the number of nodes increases from 7 to 47. Our proposal only monitors the chosen paths, so the overhead of our proposal and *Standard_SR* are nearly equal. The overhead of our proposal with 7 and 47 nodes are 3.95% and 26.71%, respectively. The overhead of *Max_QoE* is nearly three times the overhead of our proposal for 47 nodes. In other words, our proposal reduces up to 64.39% of overhead in comparison with *Max_QoE*. Although *Max_QoE* can achieve a high MOS score in the majority of episodes, it requires much resource consumption in terms of CPU usage and overhead. Consequently, it is not appropriate for a large-scale network. In contrast, the MOS score of our proposal is nearly equal or higher than the figure for *Max_QoE* in considered scenarios, but it requires less resource consumption in comparison with *Max_QoE*. Some important results are summarized as in Tables 5.3 and 5.4. Table 5.3 indicates important results related to the median MOS, 95% confidence interval (CI) of MOS and the average optimal MOS score, which is the average value of the optimal MOS score in each 50 episode of three SR mechanisms. Table 5.4 illustrates the essential results in term of the average overhead against the number of nodes of our proposal and *Max_QoE*. CI gives an estimated interval for an unknown population parameter. It is associated with a confidence level, representing a probability which the estimated interval includes a true value of the parameter. Note that 95% CI is computed at the 95% CI containing the parameter. It is calculated by the following equation:

$$CI = \overline{x} + z^{\star} \cdot \frac{\sigma}{\sqrt{n}} \qquad [5.6]$$

where $\overline{x}$ is mean of MOS, σ is its standard deviation, n is number of samples and $z^{\star}$ is 1.96 for 95% of CI.

Scenarios	Mechanisms	Average	Improvement of proposal (%)	Median	95% Confidence Interval
Perfect	Proposal	4.47	-	4.53	4.47±0.047
	Standard_SR	4.46	Nearly equal	4.52	4.46±0.056
	Max_QoE	4.47	Equal	4.53	4.47±0.047
Delay	Proposal	3.86	-	3.82	3.86±0.1
	Standard_SR	2.76	139.8	2.80	2.76±0.05
	Max_QoE	3.93	Nearly equal	4.01	3.93±0.062
Loss	Proposal	3.9	-	3.93	3.9±0.089
	Standard_SR	2.26	172.6	2.11	2.26±0.023
	Max_QoE	2.86	136.4	3.11	2.86±0.154
Delay+ Loss	Proposal	3.76	-	3.70	3.76±0.078
	Standard_SR	2.7	139.3	2.40	2.7±0.04
	Max_QoE	3.83	Nearly equal	3.72	3.83±0.024

Table 5.3. *Summarization of average optimal MOS, median and 95% confidence interval of MOS in the SR mechanisms*

Mechanisms	**Average overhead (%)**			
	7	17	32	47
Max_QoE	8.33	24.99	50	75
Proposal	3.95	9.64	18.18	26.71
Improvement of proposal (%)	**52.58**	**61.43**	**63.64**	**64.39**

Table 5.4. *Summarization of average overhead in the SR mechanisms*

In Table 5.3, 95% CI of the new SR mechanism is wider than the figure for *Max_QoE* in four scenarios. The reason is that the proposal needs to explore the routing paths to select the appropriate one, so its MOS fluctuates more frequently than the figure for *Max_QoE*.

5.5. Conclusion

Segment routing needs to be performed more adaptively to avoid network problems (e.g. congested links, etc.) and meet different service-level agreement requirements. To cope with these demands, we present a novel

SDN-based adaptive segment routing framework for network operators in the context of encrypted traffic. Our proposal is developed on the SDN controller that can be integrated into networks supporting virtualized architectures related to SDN. The novel segment routing mechanism implements different routing policies corresponding to various applications and meets strict service-level agreement requirements. Moreover, the appropriate routing path is selected according to reinforcement learning policy and the feedback of the network environment (QoE). The experimental results show that the new SR mechanism using reinforcement learning outperforms the standard SR mechanism in terms of QoE and reduces up to 64.39% of overhead in comparison to *Max_QoE* mechanism.

Segment list is one of the important factors of the segment routing mechanism, so path encoding algorithm needs to be investigated thoroughly to optimize the performance of the segment routing mechanism. After detecting the network problems and implementing the adaptive segment routing mechanism to reduce its influences, the root causes of the issues needs to be considered to deal with it definitely. Therefore, the root cause analysis mechanism will be investigated in our future work.

After reducing the negative impacts of anomalies in the network (e.g. high latency, high loss, etc.) with the temporary remediation module, we need to identify the root causes of anomalies and solve it completely. In the following chapter, we present an use-case for the root cause analysis and definitive remediation to identify the root cause of congestion and address it definitively.

6

Root Cause Analysis and Definitive Remediation

"The spread of computers and the Internet"

Marc Andreessen

Chapter 6 presents a use-case for the root cause analysis and definitive remediation. Concretely, we present a root cause analysis mechanism using machine learning (ML) to identify the root causes of congestion, and then we implement an adaptive congestion control algorithm in the definitive module to solve it completely. These modules are described in detail in Figure 2.2.

6.1. Root cause analysis: machine learning based root cause analysis for SDN network

6.1.1. *Introduction*

Root cause analysis refers to a process of identifying and delimiting elements leading to anomalies. There are three main objectives in the root cause analysis (Mdini 2019):

– The first objective is to identify which network problems result in the anomalies (e.g. link failure, switch failure, etc.). If a network problem occurs, it will be classified to return the type of the problem (Fortes et al. 2015).

– The second objective is to identify events that lead to the anomalies. Event logs of network elements can be used to study their causalities and identify anomalous events (Nagaraj et al. 2012).

– The final objective is to localize network elements that result in the anomalies. Performance evaluation at network elements can be used to identify the anomalous elements (Dimopoulos et al. 2015).

The root cause analysis can return the type of network problems, anomalous events and anomalous network elements. In this chapter, we focus on identifying the type of problems that results in network anomalies. We assume that one problem results in anomalies in the network. When many problems simultaneously lead to anomalies in the network, network administrators need to troubleshoot problems manually to solve them completely.

In the past, administrators have been able to troubleshoot network problems (e.g. using *ping*, *traceroute*, etc.) and solve them manually. However, this is not effective because of the huge number of network devices and human intervention. Consequently, root cause analysis has been studied by the research community. Conventional root cause analysis (e.g. knowledge-based mechanisms, etc.) (Hochenbaum et al. 2017; Chen et al. 2019) identifies the root causes using rules and policies with a specific threshold. However, an effective threshold identification is sometimes complicated. The emerging of ML and deep learning (DL) is a potential solution to solve this drawback. In fact, root cause analysis using ML and DL is studied by the research community to troubleshoot the network anomalies (Hong et al. 2020; Cid-Fuentes et al. 2018). Hong et al. (2020) proposed a root cause analysis method using ML to identify the problems related to service-level agreement (SLA) violations for virtual network management. Similarly, Kawasaki et al. (2020) proposed an ML-based fault classification to identify three kinds of problems, including node-down, interface-down and CPU overload. However, these studies have not considered a balance between the accuracy of ML algorithms and the processing time to select the appropriate algorithm yet.

In this section, we implement a root cause analysis (RCA) mechanism using ML and time-series network parameters to identify network problems leading to anomalies in the SDN environment. In the root cause analysis, the balance between the accuracy and processing time of different ML algorithms (e.g. random forest, gradient boosting, convolution neural network, etc.) is considered to select the appropriate algorithm. Unlike the existing studies that select the ML algorithm with the best performance, we select the ML algorithm with good performance and low time complexity to improve processing time when there is vast network traffic.

The remainder of this section is organized as follows. Section 6.1.2 presents related work on root cause analysis. The ML-based RCA mechanism is discussed in section 6.1.3. Section 6.1.4 describes the experimental results. This section concludes with section 6.1.5, which highlights our future work.

6.1.2. *Root cause analysis mechanisms*

This section presents related work on root cause analysis using knowledge-based mechanisms, causality/dependency graphs and ML-based mechanisms.

6.1.2.1. *Knowledge-based mechanism*

This mechanism (called an expert system) requires prior knowledge about anomalies to build rules to identify the types of problems and localize their locations. A rule is presented in a form as follows: *if* <symptoms> *then* <root cause>. Zhou et al. (2018) proposed BigRoots, a rule-based root cause analysis mechanism in big data systems. BigRoots uses four kinds of features, including discrete, numerical, resource and time features. The objective is to identify the problems related to CPU, I/O (input/output) and the network. Hochenbaum et al. (2017) proposed a root cause analysis in cloud infrastructure using statistical rules. This approach uses system, application and core driver metrics to localize the location of anomalies.

Despite a high accuracy, the knowledge-based mechanism contains two main disadvantages. First, it cannot identify unknown problems that are not in the rule datasets. Consequently, this approach requires updating rule datasets frequently when there are new network problems. This results in a high maintenance cost. Second, searching in a large rule dataset is costly and time consuming.

6.1.2.2. *Causality/dependency graph*

A causality graph is a graph that shows a relationship between network problems and symptoms. The objective of the causality graph is to represent the initial problems (root causes) and symptoms, and the model complex chains of intermediate problems related to these symptoms. For example, the Bayesian network is an instance of the causality graph. The Bayesian network is a directed acyclic graph (DAG) representing cause-and-effect relationships between random variables. A node in the DAG is a random variable (e.g. network elements, problems, etc.), whereas an edge is a causality between two connected nodes. Conditional probabilities express the strength of these relationships. This approach requires a deep knowledge of cause-and-effect relationships between network problems and symptoms.

Bennacer et al. (2012) proposed a root cause analysis approach using a Bayesian network to localize locations of anomalous elements. Benayas et al. (2018) proposed a root cause analysis using a Bayesian network in Big Data infrastructures. The objective is to infer 10 kinds of problems (e.g. changing flow priorities, modifying in-port rules, etc.) because of the Bayesian network and network parameters (e.g. change in a number of hosts, change in time-out, etc.).

The Bayesian network has two main drawbacks. First, it depends on reliable prior knowledge to build DAG models, so it requires deep knowledge about problems and symptoms. Second, the complexity of inference in the Bayesian network increases significantly with a huge number of nodes, so it is not effective with large-scale networks (Bennacer et al. 2012).

6.1.2.3. *ML-based mechanism*

Each network problem has a specific behavior that leads to a particular range of parameters (e.g. network parameters, resource parameters, etc.). ML-based mechanisms analyze these parameters to identify the root cause of anomalies based on classifications. In fact, many studies considered ML-based root cause analysis to detect which problems lead to anomalies in the network.

Qiu et al. (2018) proposed a novel method using ML for identifying the root cause of anomalies in NFV infrastructure. This approach collects and analyzes the features of CPU consumption, disk I/O and memory consumption using ML algorithms including neural networks, neural networks + support vector machines (SVM), K-nearest neighbors, linear SVM, radial basis function SVM, decision trees and random forest. This approach aims to identify the problems related to CPU, memory and I/O.

Kawasaki et al. (2020) proposed an ML-based fault classification to analyze the root cause of failures in the NFV environment. This approach collects 41 features from the network environment and analyzes these features using ML algorithms (e.g. random forest, support vector machine, etc.) to identify three kinds of problems including node-down, interface-down and CPU overload.

Similarly, Hong et al. (2020) proposed an anomaly detection method using ML to identify the problems related to resource usage and SLA violations. This method identifies the problems (e.g. high CPU utilization, lack of memory, etc.) based on 25 features and ML algorithms including distributed random forest, gradient boosting, extreme gradient boost and DL.

These studies have not considered the balance between the accuracy and the time complexity of ML algorithms yet. ML-based mechanisms have two main drawbacks. First, they require a large labeled dataset to train a root cause analysis model. Second, they cannot effectively identify unknown problems, which are not in training datasets.

6.1.3. *ML-based RCA mechanism*

This section presents the root cause analysis using ML to identify the root cause of anomalies in the network.

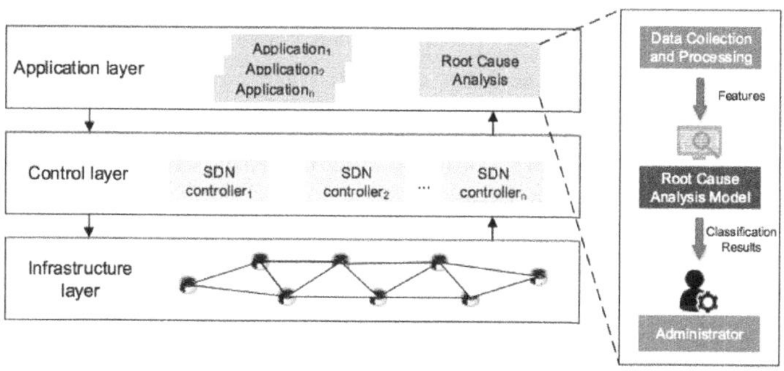

Figure 6.1. *Overall architecture of ML-based RCA in SDN environment. For a color version of this figure, see www.iste.co.uk/tong/troubleshooting.zip*

The overall architecture of ML-based RCA in the SDN environment is depicted as in Figure 6.1. First, network traffic is collected from the infrastructure layer and processed in the data collection and processing module to extract network features. Then, these features are analyzed in the root cause analysis model to identify the type of problems resulting in anomalies in the network. Finally, classification results are notified to the administrators to operate the network effectively.

6.1.3.1. *Data collection and processing*

The data collection and processing module aims to collect and extract network features corresponding to the network problems for the RCA model. According to a report of network operators (Zeng et al. 2012b), we consider three kinds of problems, including link failure, switch failure and buffer overload. These problems occur unpredictably in the network, so building troubleshooting datasets related to these problems is not accessible. Therefore, we use a fault injection technique to generate network problems to the network as in existing studies (Cid-Fuentes et al. 2018; Hong et al. 2020). Link failure is simulated by generating latency in links, while switch failure is simulated by generating rule failures in the switches to bring packet loss in links. Buffer overload is simulated by generating a vast amount of traffic surpassing the capacity of links.

These network problems are simulated in two scenarios including static and dynamic networks. In static networks, the delay and loss in the links are tied to a specific value. In dynamic networks, the link state is changed between normal and error state according to the Gilbert–Elliot (GE) model (Bildea et al. 2015) following a probability value after a given time (400 s). In this model, the probability of changing from normal to error state and from error to normal state are p and r, respectively. When a link is in the error state, the loss and delay change as presented in Table 6.1.

Root causes	Static Network	Dynamic Network
Link failure	Delay: 125ms	Delay: [25, 50, 75, 100, 125ms]
Switch failure	Loss: 50%	Loss: [10, 20, 30, 40, 50%]
Buffer overload	Bandwidth: 10 mbps, Sending rate: 60-100 mbps	Bandwidth: 10 mbps, Sending rate: 60-100 mbps

Table 6.1. *Considered network conditions*

In each network condition, we collect and extract nine features related to network and standard parameters according to existing studies (Li et al. 2018; Amour et al. 2019) and API *PortStatistics* (Team 2021). These parameters contain latency, packet loss, link utilization, number of packets sent, number of packets received, number of bytes sent, number of bytes received, number of flow entries in switch and quality of experience (QoE). Then, these parameters on each link are aggregated to calculate the path's features to identify different kinds of network problems. After that, these features are normalized into a value between 0 and 1 according to max normalization (Singh and Singh 2020). In RCA, we consider time-series features, and a sample aggregates features of 10 consecutive time steps to identify the network problems. An input sample is a matrix 10×9 (number of time steps $\times$ number of features).

We built troubleshooting datasets in static and dynamic networks. The number of network states corresponding to each problem is depicted as presented in Table 6.2.

Root causes	Static Network	Dynamic Network
Buffer overload	6900	6900
Link failure	9995	19710
Switch failure	7560	17565

Table 6.2. *Troubleshooting datasets*

6.1.3.2. *ML-based RCA method*

The features collected from the data collection and processing module are analyzed in an ML-based RCA method to identify different kinds of problems. The existence of problems leads to a fluctuation of these features. According to ML algorithms, this fluctuation can be detected to identify the root causes of anomalies. The detail of this approach is shown in Figure 6.2.

The two main phases in this method are training and testing. In the training phase, the network features will be analyzed according to an ML algorithm to obtain a pre-trained RCA model for the testing phase. The RCA model aims to infer problems in the network. There are different kinds of ML algorithms for the RCA. Therefore, we evaluate several algorithms to select the appropriate one and balance between accuracy and time complexity. The considered ML algorithms include SVM (Ahmad et al. 2018), bagging (Niranjan et al. 2017), random forest (RF) (Ahmad et al. 2018), Adaboost (Rehman Javed et al. 2020), gradient boosting (Frery et al. 2017) and convolutional neural networks (CNN) (Kwon et al. 2018). Regarding the CNN model, we use two convolutional 1D (one dimensional) layers, a max Pooling 1D, a flatter layer and three fully connected layers. For the other algorithms, we adjust hyperparameters such as the number of estimators, depth of tree, etc. These configurations are chosen according to existing studies (Hong et al. 2020) to optimize the performance of ML algorithms.

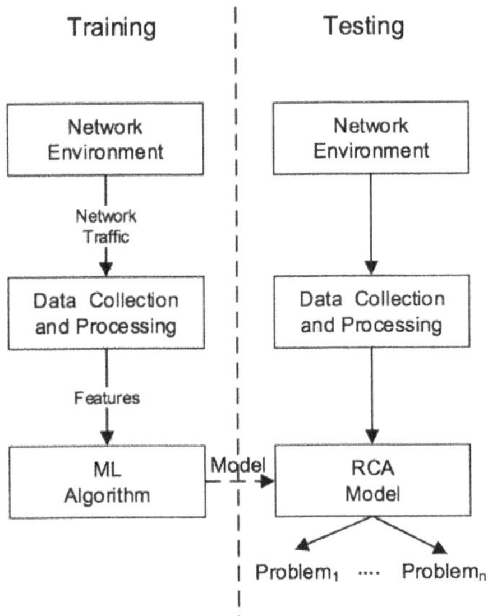

Figure 6.2. *The ML-based RCA method*

6.1.4. *Experimental results*

6.1.4.1. *Experimental setup*

In the experiments, *mininet* (Dholakiya et al. 2021) is used to generate network topology that contains five spine nodes and two leaf nodes. The delay and loss on each link is configured because of *mininet*. In the GE model,

p and r are set to 0.81 and 0.07, respectively. as in Bildea et al. (2015). The ML algorithms are implemented according to the *scikit-learn* library (Pedregosa et al. 2011) in Python. The training, validation and testing phases comprise 76%, 4% and 20% of the troubleshooting dataset, respectively. The datasets and source code of the ML-based RCA mechanism are published in Tong (2021b).

The performance of the ML algorithms is evaluated using performance metrics including precision, recall and F1-score (Sokolova and Lapalme 2009). Precision is the percentage of relevant flows that are retrieved, while recall is the percentage of retrieved flows that are relevant. The F1-score (F-measure) represents a harmonic mean between precision and recall. The details of these parameters are depicted in equations [6.1]–[6.3]. There are two ways to evaluate the quality of the overall classification, including micro-averaging and macro-averaging value. In macro-averaging, a metric is averaged over all classes that are treated equally, whereas micro-averaging is based on the cumulative True Positive (TP), False Positive (FP), True Negative (TN) and False Negative (FN) of the dataset.

$$Precision = \frac{TP}{TP + FP}. \quad [6.1]$$

$$Recall = \frac{TP}{TP + FN}. \quad [6.2]$$

$$F1 - score = \frac{2}{1/Recall + 1/Precision}. \quad [6.3]$$

6.1.4.2. *Performance analysis*

6.1.4.2.1. Dataset in a static network

There are two datasets in static and dynamic networks. First, we evaluate the balance between the accuracy and the time complexity of ML algorithms to select the appropriate one with the dataset in a static network. The performance metrics of these algorithms are described in Table 6.3. Ensemble learning is a model that makes predictions according to different models. There are two popular ensemble methods: bagging and boosting. Bagging trains individual models on a different subset of datasets in parallel and aggregate to improve the performance. In contrast, boosting trains individual models in a sequence, and each model learns mistakes from the previous one to enhance the performance. In this section, bagging uses SVM as a base classifier, so its F1-score is higher than the figure for SVM. RF is the ML algorithm using the bagging ensemble method and decision tree as the individual model, so its F1-score (approximately 96.3%) is higher than the figure for SVM and bagging.

Class	Precision						Nb Samples
	SVM	Bagging	RF	Adaboost	CNN	Gradient Boosting	
Buffer Overload	0.8040	0.8525	0.9380	0.9637	0.9593	0.9675	1380
Link Failure	0.9007	0.9802	0.9715	0.9670	0.9658	0.9764	1999
Switch Failure	0.9739	0.9875	0.9776	0.9617	0.9834	0.9791	1512
Micro	**0.8920**	0.9403	**0.9634**	0.9644	**0.9691**	**0.9746**	4891
Macro	**0.8928**	0.9401	**0.9624**	0.9641	**0.9695**	**0.9743**	4891
Class	**Recall**						**Nb Samples**
	SVM	**Bagging**	**RF**	**Adaboost**	**CNN**	**Gradient Boosting**	
Buffer Overload	0.8797	0.9928	0.9870	0.9819	0.9899	0.9928	1380
Link Failure	0.8844	0.9415	0.9560	0.9540	0.9745	0.9735	1999
Switch Failure	0.9134	0.8909	0.9517	0.9623	0.9431	0.9597	1512
Micro	**0.8920**	0.9403	**0.9634**	0.9644	**0.9691**	**0.9746**	4891
Macro	**0.8925**	0.9417	**0.9649**	0.9661	**0.9692**	**0.9753**	4891
Class	**F1-score**						**Nb Samples**
	SVM	**Bagging**	**RF**	**Adaboost**	**CNN**	**Gradient Boosting**	
Buffer Overload	0.8401	0.9173	0.9619	0.9727	0.9743	0.9928	1380
Link Failure	0.8925	0.9604	0.9637	0.9605	0.9701	0.9735	1999
Switch Failure	0.9427	0.9367	0.9645	0.9620	0.9629	0.9597	1512
Micro	**0.8920**	0.9403	**0.9634**	0.9644	**0.9691**	**0.9746**	4891
Macro	**0.8918**	0.9382	**0.9633**	0.9651	**0.9691**	**0.9753**	4891

Table 6.3. *Performance metrics of the considered ML algorithms for the dataset in a static network*

Adaboost and gradient boosting are the ML algorithms that use the boosting ensemble method and decision tree as the individual model. The objective of the boosting ensemble method is to learn from the mistakes. Adaboost learns from the mistakes by increasing the weight of misclassified samples, while gradient boosting uses the gradient instead of adjusting the weight. Hence, gradient boosting is more flexible than Adaboost. As a result, the micro and macro F1-score of gradient boosting is slightly higher than the figure for Adaboost. Moreover, the F1-score of CNN is nearly similar to gradient boosting, with 97%.

Despite the good performance, Adaboost, CNN and gradient boosting require a lot of processing time. The time complexity of the considered ML algorithms is depicted as in Table 6.4. This table describes the average processing time for a sample in the training and testing phase. The training and testing time of RF are the lowest in the considered ML algorithms with 0.03084 and 0.00143 ms, respectively. The testing time of Adaboost, CNN and gradient boosting are nearly 10, 18 and 4 times the testing time of RF, while the difference in the F1-score between these algorithms is less than 1%. Therefore, we consider RF as a ML algorithm for RCA.

Algorithms	**Training Time (ms)**	**Testing Time (ms)**
SVM	**1.21472**	**0.49883**
Bagging	37.9581	35.8988
RF	**0.03084**	**0.00143**
Adaboost	0.36427	0.01378
CNN	**26.044**	**0.02636**
Gradient Boosting	**1.48182**	**0.00555**

Table 6.4. *Time complexity of ML algorithms in RCA for the dataset in a static network*

We use nine features to identify the root cause of anomalies, but some of these features are ineffective for the RCA. Consequently, we implement a feature selection method (wrapper method) to identify the appropriate feature set. The wrapper method (Al-Tashi et al. 2020) evaluates all possible feature sets based on a specific ML algorithm and selects the feature set with the highest accuracy. RF is considered as a learning algorithm in the wrapper method due to the balance between its accuracy and time complexity.

Figure 6.3 illustrates the accuracy against the number of features in the feature selection method. The feature set with seven features indicates better results than the others, so this feature set is considered in the RCA. This feature set contains latency, link utilization, number of packets received, number of bytes sent, number of bytes received, number of flow entries in switch and QoE. Table 6.5 shows the F1-score of the selected feature set (Feature Set 2) in comparison with the feature set with all nine features (Feature Set 1).

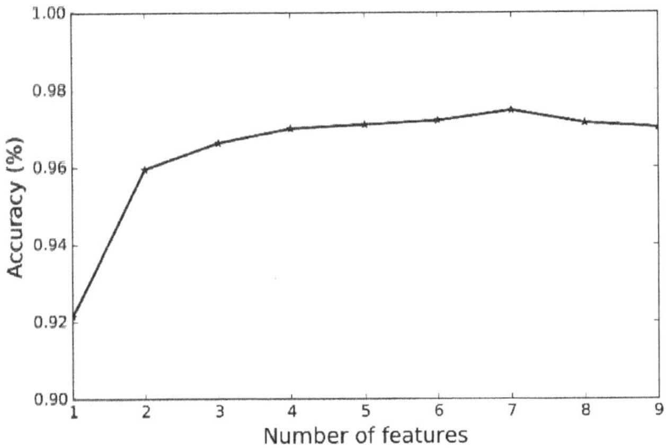

Figure 6.3. *The accuracy against the number of features in the feature selection method. For a color version of this figure, see www.iste.co.uk/tong/troubleshooting.zip*

Class	F1-score		Nb Samples
	Feature Set 1	Feature Set 2	
Buffer Overload	0.9619	0.9748	1380
Link Failure	0.9637	0.9710	1999
Switch Failure	0.9645	0.9674	1512
Micro	**0.9634**	**0.9710**	4891
Macro	**0.9633**	**0.9710**	4891

Table 6.5. *F1-score of two feature sets in the RCA for the dataset in a static network*

6.1.4.2.2. Dataset in a dynamic network

Tables 6.6 and 6.7 illustrate the performance metrics and the time complexity of ML algorithms with the selected feature set in the RCA for a dynamic network. Similar to the dataset in a static network, the micro and macro F1-scores of gradient boosting are the highest (approximately 95%), while the testing time of RF is the lowest in the considered ML algorithms for the dataset in a dynamic network. The difference of the F1-score between RF and gradient boosting is approximately 1%, so the RF is considered as an ML algorithm in RCA for the dataset in a dynamic network. The RCA can achieve good results in the dynamic network with over 93% for the micro and macro precision, recall and F1-score. The micro and macro F1-score in the dynamic network are approximately 94%, and reduce about 3% compared to the figure for the static network. The reason for this is that the error status, as well as the delay and loss value change after a given time in the dynamic network. This leads to less fluctuation in the network features, and therefore reduces these performance metrics.

6.1.5. *Conclusion*

In this section, we presented a root cause analysis approach using ML and time-series network parameters to identify the root cause of anomalies (the type of problems leading to the anomalies) in the SDN environment. There are various ML algorithms, so we considered the balance between accuracy and time complexity to select the appropriate algorithm. The experimental results illustrate that the precision, recall and F1-score of the root cause analysis approach are approximately 97% in the static network.

In the future, we will extend the troubleshooting datasets to consider more network problems. There is a massive amount of useful resources (e.g. system logs, etc.) in addition to the network parameters to enhance the performance of root cause analysis. Therefore, system logs, as well as data-processing applications (e.g. Hadoop, Splunk, etc.) will be considered to improve the performance of root cause analysis.

Class	Precision						Nb Samples
	SVM	Bagging	RF	Adaboost	CNN	Gradient Boosting	
Buffer Overload	0.6710	0.7813	0.9191	0.9201	0.9040	0.9535	1380
Link Failure	0.7797	0.8669	0.9471	0.9244	0.9483	0.9487	3942
Switch Failure	0.8492	0.8554	0.9451	0.9326	0.9292	0.9578	3513
Micro	**0.7914**	0.8491	**0.9418**	0.9269	**0.9333**	**0.9530**	8835
Macro	**0.7666**	0.8345	**0.9371**	0.9257	**0.9272**	**0.9534**	8835

Class	Recall						Nb Samples
	SVM	Bagging	RF	Adaboost	CNN	Gradient Boosting	
Buffer Overload	0.5659	0.7739	0.9464	0.9341	0.9688	0.9667	1380
Link Failure	0.8630	0.8757	0.9500	0.9434	0.9302	0.9619	3942
Switch Failure	0.7996	0.8488	0.9308	0.9055	0.9229	0.9377	3513
Micro	**0.7914**	0.8491	**0.9418**	0.9269	**0.9333**	**0.9530**	8835
Macro	**0.7428**	0.8328	**0.9424**	0.9277	**0.9406**	**0.9554**	8835

Class	F1-score						Nb Samples
	SVM	Bagging	RF	Adaboost	CNN	Gradient Boosting	
Buffer Overload	0.6140	0.7776	0.9325	0.9270	0.9353	0.9601	1380
Link Failure	0.8192	0.8712	0.9486	0.9338	0.9392	0.9553	3942
Switch Failure	0.8236	0.8521	0.9379	0.9188	0.926	0.9476	3513
Micro	**0.7914**	0.8491	**0.9418**	0.9269	**0.9333**	**0.9530**	8835
Macro	**0.7523**	0.8336	**0.9397**	0.9266	**0.9335**	**0.9543**	8835

Table 6.6. *Performance metrics of the considered ML algorithms for the dataset in a dynamic network*

Algorithms	Training Time (ms)	Testing Time (ms)
SVM	**3.70088**	**1.43646**
Bagging	81.4241	67.4686
RF	**0.38460**	**0.00587**
Adaboost	0.35830	0.01368
CNN	**26.1040**	**0.02455**
Gradient Boosting	**1.40897**	**0.01087**

Table 6.7. *Time complexity of ML algorithms in RCA for the dataset in a dynamic network*

6.2. Definitive remediation: adaptive QUIC BBR algorithm using reinforcement learning for dynamic networks

6.2.1. *Introduction*

Today, the Internet continues to grow rapidly, leading to many hindrances for service providers such as latency optimization, congestion, etc. Growth in latency-sensitive web service leads to various strict latency requirements from the end-user's side. Moreover, the Internet is changing from insecure to secure traffic rapidly, which leads to more latency in data transmission (e.g. additional latency due to the TLS handshake, etc.). The emergence of Quick UDP Internet Connections (QUIC) (Langley et al. 2017), a new transport layer network protocol developed by Google, is a potential solution to improve latency in the network. In early 2021, QUIC was standardized by the Internet Engineering Task Force (IETF) to offer opportunities to reduce latency in connection establishment and prevent HoL blocking (head-of-line blocking) in comparison with TCP+TLS. QUIC accounts for over 30% of Google's total egress traffic and approximately 7% of global Internet traffic (Langley et al. 2017). According to recent statistics, global Internet traffic is expected to reach 333 exabytes (EB) per month by 2022 compared with 100 exabytes per month in 2017 (Statista 2022). The rapid growth in global Internet traffic has created a lot of pressure for network infrastructure and results in congestion in the network, particularly during the Covid-19 pandemic. Therefore, congestion control plays an essential role in the network. In fact, congestion control is considered by much existing work to prevent congestion in the network (Dong et al. 2018; Wu et al. 2020).

However, Google implemented a static congestion control mechanism for QUIC in its infancy.

In the early stage, QUIC used Cubic for congestion control like in TCP. Cubic is a loss-based approach that changes a cwnd (congestion window) parameter to adjust an amount of inflight data (data sent but not yet acknowledged) in a buffer of network devices based on a loss signal. Loss happens in the network when the buffer is overloaded. Consequently, a large buffer size leads to a bufferbloat, influencing interactive applications (e.g. multiplayer online games, etc.). On the other hand, a small buffer size results in high packet loss and low throughput, impacting multimedia applications (e.g. video streaming, etc.). Moreover, using loss signal in congestion control can lead to uncertain results because non-congestion packet loss is a widespread phenomenon in the network (e.g. port flap in routers, lossy wireless links, etc.) (Dong et al. 2018).

To overcome these disadvantages, Google has developed BBR (bottleneck bandwidth and round-trip time (RTT)), a new congestion control algorithm (Hock et al. 2017). BBR is a rate-based approach that monitors delivery rate and minimal RTT to adjust sending rate. However, a static congestion control algorithm cannot be effective across different network conditions (Nie et al. 2019). After evaluating the performance of BBR and its variants across network conditions (section 6.2.4.2), we conclude on the necessity of an adaptive congestion control algorithm for dynamic networks. Therefore, in this section, we present an adaptive BBR (A-BBR) algorithm using reinforcement learning (RL) for QUIC protocol. The experimental results (section 6.2.4.2) show that A-BBR has a higher overall average reward by 18.6%, 30.76%, 13.3% and 27.5% compared to origin BBR and three other BBR variants. The fairness index of A-BBR is similar to the figure for the others (approximately 1), showing fair bandwidth sharing between the flows of these congestion control algorithms. The contributions of this work are listed as follows:

– A-BBR aims to select the appropriate policies to adaptively change the sending rate in accordance with unprecedented changes in network environments.

– A-BBR with RL allows the implementation of a real-time solution in congestion control, unlike existing studies using deep reinforcement learning (DRL) (Nie et al. 2019), which requires high resource consumption and time complexity.

– Performance of A-BBR and benchmarks are evaluated in the context of HTTP/3 (Hypertext Transfer Protocol version 3) in contrast to existing studies evaluating congestion control algorithms in the context of HTTP/2 (Wang et al. 2018). HTTP/3 offers opportunities to solve the drawbacks of HTTP/2 in terms of HoL blocking and latency reduction in connection establishment (Langley et al. 2017). To the best of our knowledge, this is the first study comparing the performance of congestion control algorithms for QUIC in the context of HTTP/3.

The remainder of this section is organized as follows. Section 6.2.2 presents related work on loss-based congestion control algorithms, rate-based congestion control algorithms and the improvement of these algorithms. The A-BBR algorithm is discussed in section 6.2.3. Section 6.2.4 describes the experimental results of A-BBR. This section concludes with section 6.2.5, which highlights future work.

6.2.2. *Congestion control mechanisms*

In this section, we present related work on congestion control algorithms such as loss-based congestion control algorithms (New Reno and Cubic), rate-based congestion control algorithms (BBR) and the improvement of rate-based congestion control algorithms. These algorithms are supported in the open-source libraries of the QUIC protocol (e.g. *lsquic*, *picoquic*, *quic-go*, *quiche*, etc.) (Marx et al. 2020).

6.2.2.1. *Loss-based congestion control*

According to related work (Patel et al. 2020; Abdeljaouad et al. 2010), there are many congestion control algorithms in the network, such as Tahoe, New Reno, Westwood, BIC, Cubic, etc. New Reno has been implemented in Windows XP, while Cubic has been deployed in Linux kernels version 2.6.19 and above from 2006 (Abdeljaouad et al. 2010). These algorithms are loss-based congestion control algorithms that adjust the sending rate based on the feedback of receivers (e.g. acknowledgment packets, etc.).

The loss-based congestion control algorithms contain three main phases: *Slow Start, Recovery* and *Congestion Avoidance*. A sender enters the *Slow Start* phase when its congestion window is less than the *ssthresh* threshold, which is initialized to an infinite value in the beginning. Then, cwnd doubles every RTT to increase the sending rate quickly in the early stage. The sender moves to the *Recovery* phase when there is a loss in the network. The loss happens when one of the following two conditions is met. First, the elapsed time from when a packet is sent is equal to or bigger than a threshold (9×RTT/8). Second, the difference between the highest packet number of acknowledged packets and the packet number of the unacknowledged packet is equal to or bigger than three. In the *Recovery* phase, the sender re-transmits lost packets and reduces the *ssthresh* threshold to the half value of the current cwnd in New Reno (or reduces ssthresh by 30% in Cubic). It sets the current cwnd to the current *ssthresh* threshold before moving to the *Congestion Avoidance* phase. The sender ends the *Recovery* phase and enters the *Congestion Avoidance* phase when packets sent during the *Recovery* phase are acknowledged. In the *Congestion Avoidance* phase, the sender increases the cwnd by one every RTT in New Reno (or increases cwnd following a cubic function every RTT in Cubic) to avoid congestion.

6.2.2.2. *Rate-based congestion control*

BBR (Hock et al. 2017) is a rate-based approach that adjusts the sending rate according to a pacing_gain and a delivery rate estimated from acknowledged packets. The objective is to determine the amount of inflight data ($D^{inflight}$) in order to utilize bandwidth-delay product bdp (available bottleneck bandwidth) effectively. This value is calculated according to delivery rate and RTT as follows: $bdp = b_r \times RTT_{min}$, where b_r is the delivery rate and RTT_{min} is the minimal RTT.

Figure 6.4 shows the delivery rate and RTT against $D^{inflight}$. Operating point *A* shows that $D^{inflight}$ is equal to the bandwidth-delay product bdp. After reaching point *A*, an increase in $D^{inflight}$ does not increase the delivery rate. If $D^{inflight}$ continues to increase, the buffer starts to be filled with excess data. When the buffer is fully filled, the operating point shifts to *B* where the inflight packets start to be dropped. BBR tries to shift the operating point toward A while loss-based approaches (e.g. Cubic, etc.) move the operating point to B, which results in unexpected delay in data transmission.

There are four main phases in BBR: *Startup, Drain, ProbeBW* and *ProbeRTT*. In the *Startup* phase, BBR increases its sending rate by using a pacing_gain of $2/ln(2)$. If there are three consecutive rounds where increasing the sending rate only leads to a small growth in the delivery rate (less than 25%), BBR exits this phase and enters the *Drain* phase. In this phase, BBR decreases its sending rate using a pacing_gain of $ln(2)/2$ until the amount of inflight data matches bdp. After that, BBR enters the *ProbeBW* phase to probe more bandwidth using a pacing_gain cycle of [1.25, 0.75, 1, 1, 1, 1, 1, 1]. BBR increases the sending rate using each pacing_gain in a duration of $\mathrm{RTT}_{\mathrm{min}}$, which is updated every 10 s in the *ProbeRTT* phase. When an increase in the sending rate leads to a growth in the delivery rate, BBR sets the delivery rate to a new sending rate. Otherwise, BBR maintains the current sending rate. BBR contains a drawback related to a static congestion policy. Concretely, it uses a fixed pacing_gain sequence to adjust the sending rate, which is not effective across network conditions.

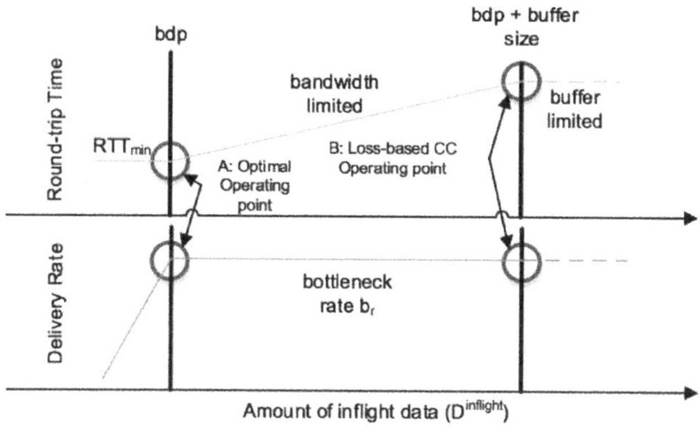

Figure 6.4. *Congestion control operating point: delivery rate and RTT against the amount of inflight data. For a color version of this figure, see www.iste.co.uk/tong/troubleshooting.zip*

6.2.2.3. *Improvement of rate-based congestion control*

In BBR, the sending rate is calculated according to the pacing_gain and the delivery rate. The pacing_gain is changed during every period of time ($\mathrm{RTT}_{\mathrm{min}}$) following the sequence of [1.25, 0.75, 1, 1, 1, 1, 1, 1]. This fixed pacing_gain sequence is suboptimal in several network environments. For example, in the wireless network, many packets will be aggregated in a single large frame to reduce overhead and increase efficiency in data transmission (Wang et al. 2019). This leads to a higher throughput and airtime consumption. Consequently, the amount of inflight data $\mathrm{D}^{\mathrm{inflight}}$ needs to be increased in order to reach bdp. Therefore, there is a need for a higher pacing_gain in order to probe more bandwidth. Wang et al. (2019) proposed

BBR+ with a novel pacing_gain sequence of [1.5, 0.5, 1.5, 0.5, 1.5, 0.5, 1.5, 0.5]. BBR+ increases the pacing_gain to 1.5 to probe more bandwidth and then reduces it to 0.5 to decrease the excess data in the buffer. Similarly, Zhang et al. (2019) presented BBR-Tsunami with a pacing_gain sequence of [1.5, 0.75, 1.25, 1.25, 1.25, 1.25, 1.25, 1.25]. According to BBR+ and BBR-Tsunami, we present another pacing_gain sequence of [2, 0.5, 1.5, 0.5, 2, 0.5, 1.5, 0.5] to probe more bandwidth in the context of high packet loss. This algorithm refers to M-BBR (modified BBR).

A static congestion control algorithm is ineffective for various network conditions, so Nie et al. (2019) proposed TCP-RL, an adaptive congestion control schema in TCP using deep reinforcement learning. TCP-RL takes into account three network parameters (throughput, RTT and loss) using deep reinforcement learning to select an appropriate congestion control algorithm. In contrast, we present an adaptive BBR algorithm in QUIC protocol for dynamic networks.

6.2.3. *Adaptive BBR algorithm*

In this section, an A-BBR algorithm is presented to perform effectively across network conditions. This algorithm is inspired by a reinforcement learning task that selects the appropriate set of pacing_gain thanks to feedback from the network environment to optimize an objective function. This section determines and formalizes the appropriate policies to solve this task.

Selecting the appropriate pacing_gain sequence is formalized as a reinforcement learning task (Figure 6.5). This is described in detail in the following:

– *Agent*: an agent is an entity in the network system executing its tasks according to a learning algorithm. In A-BBR algorithm, an agent selects the appropriate pacing_gain sequence to optimize an objective function.

– *State*: an instance of the network state that is monitored by the agent.

– *Action*: an action describes how an agent reacts to the network environment. In the A-BBR algorithm, the action is a pacing_gain sequence corresponding to each BBR variant. The considered pacing_gain sequences contain [1.25, 0.75, 1, 1, 1, 1, 1, 1] (Origin-BBR), [1.5, 0.5, 1.5, 0.5, 1.5, 0.5,

1.5, 0.5] (BBR+), [1.5, 0.75, 1.25, 1.25, 1.25, 1.25, 1.25, 1.25] (BBR-Tsunami) and [2, 0.5, 1.5, 0.5, 2, 0.5, 1.5, 0.5] (M-BBR). Changing the pacing_gain leads to the change in sending rate corresponding to various network conditions. These pacing_gain sequences are selected to probe more available bandwidth across network conditions (e.g. lossy networks, etc.).

– *Policy*: a policy is a mapping between a state and an action in the network environment.

– *Reward*: a reward is feedback from the network environment. At step t, the agent monitors a network state s and executes an action a. After that, the network state moves to $s^{'}$, and the agent receives a corresponding reward r. In several studies (Nie et al. 2019; Winstein and Balakrishnan 2013), reward is defined as follows: $r = \log \frac{Throughput}{RTT}$. However, this reward function leads to a small difference in the performance of BBR with considered pacing_gain. Therefore, reward is defined as follows: $r = \frac{Throughput}{RTT}$. The reward is normalized into a value between 0 and 1 according to the min-max normalization function (Singh and Singh 2020), to facilitate a comparison between the congestion control algorithms.

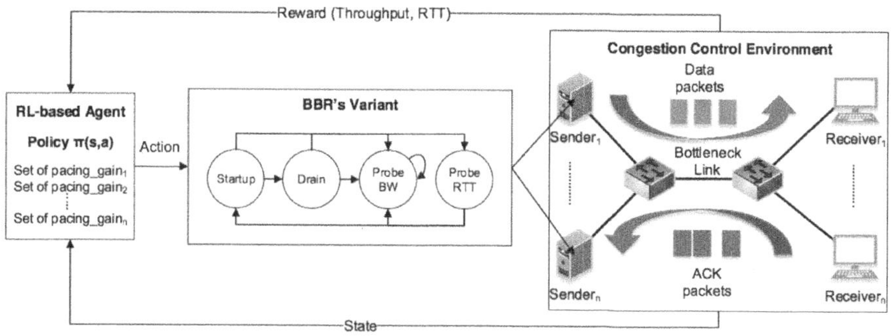

Figure 6.5. *Adaptive BBR algorithm. For a color version of this figure, see www.iste.co.uk/tong/troubleshooting.zip*

The objective of RL is to optimize the accumulative reward O_f (equation [6.4]):

$$O_f = Max\ E[\sum_{t=0}^{\infty} \gamma_t \times r_t]. \qquad [6.4]$$

where $\gamma_t \in [0, 1]$ is a discount factor.

There are two approaches in RL: model-based and model-free. The first approach learns an environment model to optimize the optimality, while the second one learns its policies without prior information in the network environment. In this work, the model-free approach is considered because it requires less storage cost and dependence on accuracy of initial information than the model-based approach (Mammeri 2019). The quality of action a is evaluated via Q-value $Q(s,a)$, which is updated at every step as in equation [6.5]:

$$Q(s,a) = (1-\alpha)Q(s,a) + \alpha[r + \gamma \max_{a'} Q(s',a')] \qquad [6.5]$$

where α is a learning rate and γ is a discount factor.

In this approach, $\epsilon - greedy$ (Tran et al. 2019) is considered as a selection algorithm to balance between exploration and exploitation phases. Concretely, the agent chooses the pacing_gain sequence with the highest Q-value with a probability of (1-ϵ). Otherwise, the agent chooses the pacing_gain sequence randomly. Then, the ϵ value is reduced at every step according to ϵ_decay, until it drops to ϵ_min.

The time complexity of the reinforcement learning algorithm is $O(m \times N)$, where N is the number of actions and m is the size of state space (Jin et al. 2020).

6.2.4. *Experimental results*

6.2.4.1. *Experimental setup*

The testbed is implemented with dumbbell topology (Figure 6.5) in a similar way as in Hock et al. (2017). *mininet* (Dholakiya et al. 2021) is used to emulate network topology and change network states in the bottleneck link. This link is set with a bandwidth of 10 mbps, a RTT of [50, 200, 100 ms] and a loss of [30, 10, 0, 20, 1, 5%] to create a variety of network conditions. Therefore, there are 18 considered network conditions corresponding to the change of loss and RTT. The number of senders and receivers is set from one to three. In the experiments, a file with a size of 40 MB is sent from servers to clients. On the sender's side, the proposal (A-BBR) and benchmarks (Cubic, Origin-BBR, BBR+, BBR-Tsunami and M-BBR) are implemented according to customizing the library *lsquic* (Dmitri and George 2021) to provide QUIC over HTTP/3. On the receiver's side, Google Chrome is used to communicate with senders via QUIC over HTTP/3. On the reinforcement learning, ϵ is initiated to 1. ϵ_decay, ϵ_min, α and γ are set to 0.95, 0.3, 0.7 and 1, respectively. These parameters are chosen according to the experiments.

6.2.4.2. *Performance analysis*

Figure 6.6 illustrates the number of network conditions in which each congestion control algorithm obtains the best performance (largest reward). There are four BBR variants corresponding to four pacing_gain sequences (described in section 6.2.3). This figure indicates that a specific congestion control algorithm cannot perform well across various network conditions. In the considered network conditions, Origin-BBR performs well overall, but it is only effective in 11 network conditions. In contrast, M-BBR performs best in none of these network conditions. This confirms the necessity of the adaptive BBR algorithm for QUIC in dynamic networks.

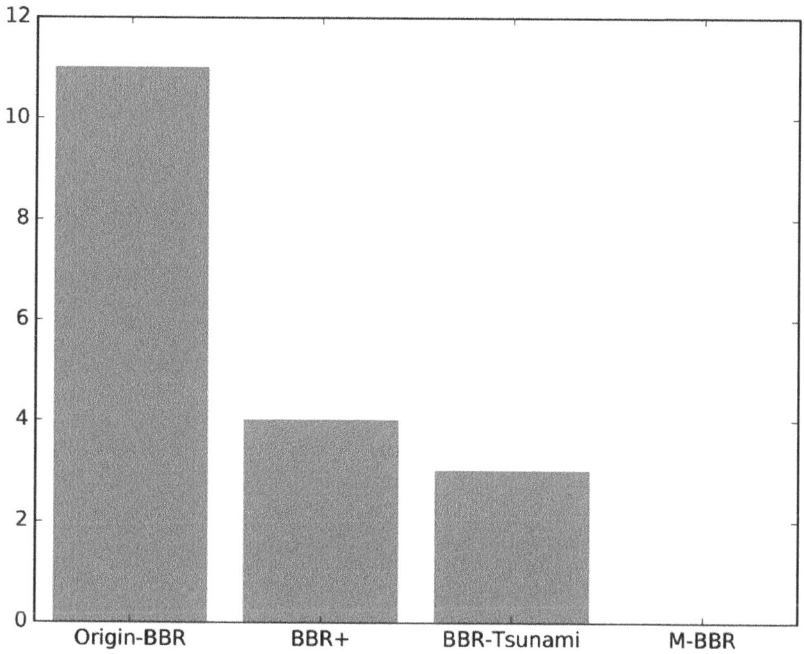

Figure 6.6. *Number of network conditions in which each congestion control algorithm obtains the best performance. For a color version of this figure, see www.iste.co.uk/tong/troubleshooting.zip*

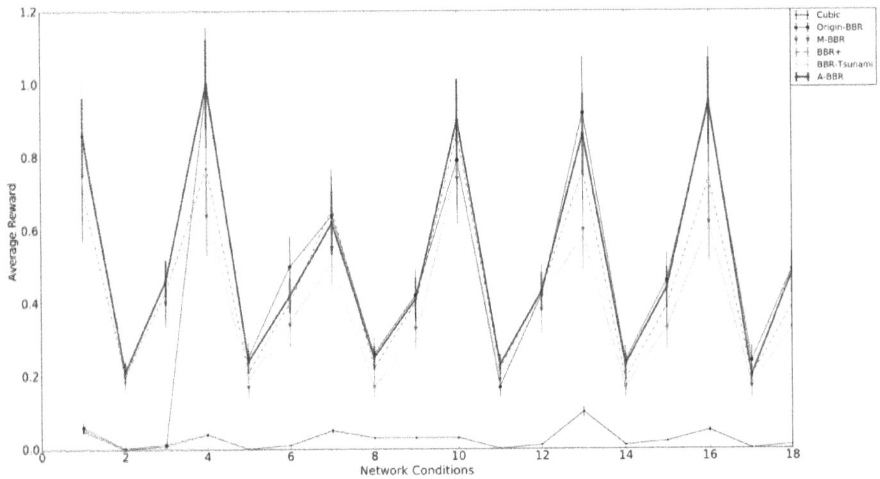

a) Average reward and standard deviation in each considered network condition

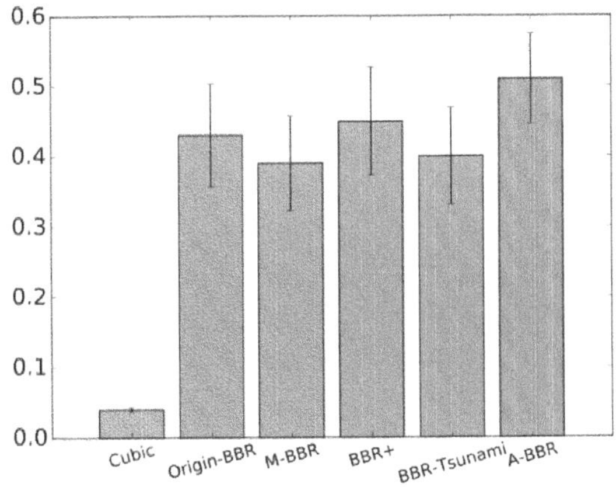

b) Overall average reward and standard deviation in all considered network conditions

Figure 6.7. *Average reward and standard deviation of A-BBR and benchmarks. For a color version of this figure, see www.iste.co.uk/tong/troubleshooting.zip*

In reality, network conditions are changed over time. Consequently, we evaluate the performance of A-BBR and benchmarks in the considered network conditions. Figure 6.7(a) shows their standard deviations and average rewards in these network conditions. A-BBR can achieve good results in network conditions that will be changed every 100 time steps. The change in these network conditions leads to the change of RTT and loss in the network. Consequently, this leads to the change of throughput, resulting in the fluctuation of reward in these network conditions. According to RL, A-BBR can learn to keep the same pacing_gain sequence or select other sequences to obtain good results and optimize the accumulative reward. Loss is set to 30%, and RTT varies from 50 to 200 in the first three network conditions. In these network conditions, the average reward of Origin-BBR ranges from 0.01 to 0.06. The reason for this is that Origin-BBR increases the sending rate with a pacing_gain of 1.25 for a duration (RTT_{min}), but there is 30% of loss in this scenario. This leads to a decrease in throughput and reward against time. When Origin-BBR tries with lower pacing_gain value, its throughput and reward continue to decrease against time. Moreover, the average reward of Cubic is the lowest in the considered congestion control algorithms. In these network conditions, there is a loss in the network. Cubic reduces its cwnd according to loss signal, so there is a decrease in its throughput and average reward. A noticeable feature from Figure 6.7(a) is that the average reward of the considered congestion control algorithms reaches a peak every three network conditions. In every three network conditions, RTT is set to 50 ms in the first network condition, then increases to 200 ms in the second one and finally reduces to 100 ms in the last one. This leads to the fluctuation of average reward in these algorithms. The overall average reward and standard deviation of these congestion control algorithms obtained in all considered network conditions are illustrated in Figure 6.7(b). The overall average reward of the proposal is the highest with 0.51, while the figures for the Origin-BBR, M-BBR, BBR+, BBR-Tsunami and Cubic are 0.43, 0.39, 0.45, 0.4 and 0.04, respectively. The overall average reward of our proposal shows an improvement of 18.6%, 30.76%, 13.3% and 27.5% in comparison with the figure for Origin-BBR and three other BBR variants. The overall standard deviation of A-BBR is nearly equal to the figure for Origin-BBR, M-BBR, BBR+ and BBR-Tsunami. Their overall standard deviations range from 0.064 to 0.077. Some important results related to overall average reward and overall standard deviation are summarized in Table 6.8.

Figure 6.8 indicates the Jain's fairness of the considered congestion control algorithms in a network condition with a bandwidth of 10 Mbps, a loss of 5% and a RTT of 100 ms. In this experiment, the file with size of 40 MB is sent from three senders to three corresponding receivers. All

Algorithms	**Origin-BBR**	**M-BBR**	**BBR+**
Overall average reward	0.43	0.39	0.45
Algorithms	BBR-Tsunami	Cubic	A-BBR
Overall average reward	0.4	0.04	0.51
Algorithms	Origin-BBR	M-BBR	BBR+
Overall standard deviation	0.073	0.067	0.077
Algorithms	BBR-Tsunami	Cubic	A-BBR
Overall standard deviation	0.069	0.003	0.064

Table 6.8. *Some important results of the considered congestion control algorithms*

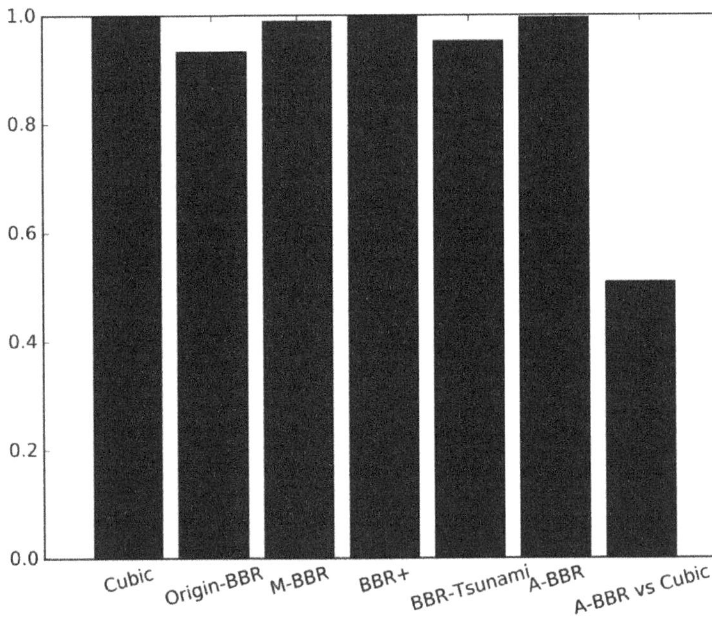

Figure 6.8. *Fairness of A-BBR and benchmarks in dynamic network conditions. For a color version of this figure, see www.iste.co.uk/tong/troubleshooting.zip*

considered congestion control algorithms can achieve a good fairness index because their values approximate to 1. The last column shows fairness between A-BBR and Cubic with a flow of A-BBR and two Cubic flows. The fairness index between A-BBR and Cubic approximates to 0.5, so there is less fair bandwidth sharing between the flows of these algorithms. With the loss of 5%, Cubic reduces its cwnd according to loss signal, while A-BBR can probe to adjust its sending rate. Therefore, the throughput of A-BBR is higher than Cubic, which leads to a low fairness index.

6.2.5. *Conclusion*

This section presented an A-BBR algorithm for QUIC to adapt to the unexpected changes in network environments, because a static congestion control algorithm (original BBR) is not effective across various network conditions. According to RL, A-BBR changes the sending rate adaptively, using the feedback from the network environments. Moreover, A-BBR and the benchmarks are evaluated in the context of HTTP/3. The experimental results show that A-BBR achieves good results related to the average reward and fairness index across the considered network conditions compared to the benchmarks.

Despite the good results, A-BBR, as well as other BBR variants contain a drawback related to the fairness with existing congestion control algorithms (e.g. QUIC Cubic, etc.). Google is developing BBR v2 that offers an opportunity to solve this drawback. In the future, we will implement the adaptive algorithm for BBR v2. We will extend this study to select appropriate congestion control algorithms (e.g. QUIC Cubic, QUIC New Reno, etc.) across various network conditions.

Conclusions and Prospects

"Software-defined networks and the maturing of the Internet"

Nick Mckeown,
IET Appleton, 2014

C.1. Summary of contributions

Nowadays, the rapid development of the Internet results in more and more multimedia services (e.g. video streaming, online game and so on) and diversified network devices such as IoT devices. The huge number of network devices not only enlarges the network infrastructures but also leads to many critical issues in the network. A downtime on cloud-based services leads to negative economic impacts (e.g. the loss from over \$30,000 to \$6,700,000) for many service providers (e.g. YouTube, Facebook, PayPal, etc.) (Cérin et al. 2013). Therefore, network troubleshooting has been increasingly concerned over the past two decades. The purpose is to detect anomalies, identify its root causes and implement remediation approaches to deal with these root causes definitively.

Although network troubleshooting is studied thoroughly by the research community, the emergence of encrypted traffic entails many questions regarding its deployment in the context of encrypted traffic. For example, Google developed Quick UDP Internet Connections (QUIC), a transport layer network protocol from 2012 to improve latency in data communication.

QUIC also provides an encryption algorithm to guarantee secure data communication in the network. In this case, from the point of view of network operators (NOs), useful information obtained in data packets is hidden, such as sequence number, acknowledgment number, payload signature, etc. This causes drawbacks for network performance monitoring approaches (e.g. QoE estimation, application identification, etc.) and intrusion detection systems. These approaches are essential components in network troubleshooting frameworks, so encrypted traffic brings certain obstacles to data collection (e.g. QoE estimation application identification, etc.) and remediation approaches (e.g. application-aware traffic engineering, payload-based intrusion detection systems, etc.) in network troubleshooting.

Moreover, computer networks nowadays have become more complex due to many network devices (e.g. IoT devices, etc.), so the processing time for root cause analysis and remediation approaches becomes larger. Therefore, network systems have negative impacts (e.g. high latency, high loss, etc.) during the root cause analysis of anomalies and remediation. Consequently, it is necessary for network troubleshooting frameworks that guarantee network availability during the root cause analysis and remediation and deal with some of the limitations encountered in encrypted traffic. Therefore, in this book, we present a novel troubleshooting framework for NOs in the context of encrypted traffic. This troubleshooting framework contains five main modules: data collection, anomaly detection, temporary remediation, root cause analysis and definitive remediation. These modules are described as follows.

– Data collection: in this module, we implement a parameter measurement module to collect troubleshooting data and build datasets in static and dynamic networks. A sample in the datasets contains nine time-series network parameters (e.g. latency, loss and etc.) and corresponding labels for three root causes: link failure, switch failure and buffer overload. From the point of view of NOs, network performance metrics (e.g. quality of experience, application classes, etc.) are hidden due to encrypted traffic. Consequently, we present a novel traffic classification approach for QUIC traffic. The information about application classes plays an important role in application-aware remediation approaches (e.g. application-aware traffic engineering, etc.). There are two main classification stages in the traffic classification approach. The first one analyzes flow-based features using a machine learning (ML) algorithm to classify the network traffic into chat, VoIP or elephant flows. Random forest is considered as an ML algorithm in this stage after evaluating the performance of several ML algorithms (e.g. support vector machine, random forest, etc.). The second classification stage analyzes packet-based features using convolutional neural network (CNN) algorithm to classify the elephant flows into video streaming, file transfer or Google Play Music. In this stage, performance is evaluated in various scenarios such as different subsets of input vector and various loss functions. The experiment results show that classification results can achieve approximately 99% for video streaming, file transfer and Google Play Music and approximately 96% for other applications such as chat and VoIP.

– Anomaly detection: after collecting troubleshooting data, this information is analyzed in anomaly detection module to detect anomalies in the network. In this book, we implement an anomaly detection approach using ML algorithm to classify network states into normal or abnormal states. The experimental results show that the precision, recall and F1-score of the anomaly detection approach achieve up to 99% in the considered datasets.

– Temporary remediation: if anomalies happen in the network, we must implement a temporary remediation module that is responsible for reducing negative impacts of problems (e.g. high latency, high loss, etc.) and guaranteeing the network availability during the root cause analysis and definitive remediation. In the temporary remediation module, we present an application-aware segment routing mechanism in the software-defined networking environment. This module is to implement a variety of routing strategies corresponding to different applications. Firstly, the temporary remediation module identifies application classes according to traffic classification. Then, this module selects an appropriate routing strategy for a specific application because of reinforcement learning (RL) algorithm and feedback from the network environment. In the RL algorithm, there are various selection algorithms (e.g. *E-greedy*, *softmax*, etc.), so we evaluate the performance of these algorithms to select appropriate ones for four scenarios: perfect scenario, delay scenario, loss scenario and scenario with both delay and loss. Therefore, the segment routing mechanism is evaluated with benchmarks (*Standard_SR* and *Max_QoE* mechanisms) related to mean opinion score (MOS) against time, average optimal MOS, median MOS, 95% confidence interval of MOS, CPU usage and control overhead. The experimental results illustrate that MOS of the proposal is nearly equal or higher than the figure for benchmarks, while it requires less CPU usage and control overhead compared to benchmarks in the considered scenarios. Concretely, the proposal reduces up to 64.39% of overhead in comparison with *Max_QoE* mechanism.

– Root cause analysis: in parallel with the temporary remediation, we implement a root cause analysis module to identify the root cause of anomalies in the network. In this module, we implement a root cause analysis approach using time-series network parameters and an ML algorithm. In the book, congestion is considered to generate anomalies in the network. The objective of root cause analysis is to identify the root cause of congestion (link failure, switch failure or buffer overload). Besides, the performance and processing time of ML algorithms (e.g. support vector machine, bagging, random forest, Adaboost, etc.) are evaluated to select an appropriate algorithm for datasets in static and dynamic networks. The experimental results show that random forest can achieve high accuracy while requiring less processing time than other algorithms. Consequently, it is considered as an ML algorithm in the root cause analysis approach. F1-score of the root cause analysis mechanism achieves up to 97% while it only requires 0.00143 ms for the testing time for the dataset in static network.

– After identifying the root cause of anomalies, we implement a definitive remediation module to solve the root cause definitively. If congestion results from link failure or switch failure, we send notifications to administrators so that they can address these root causes. Otherwise, we present an adaptive BBR (bottleneck bandwidth and round-trip time) algorithm using the RL algorithm to deal with the switch's buffer overload in the network. This approach aims to adjust sending rate at sender sides adaptively for various network conditions. The performance of the congestion control algorithm is evaluated with benchmarks in the context of HTTP/3 in contrast to HTTP/2 in existing studies. The experimental results show that this congestion control algorithm achieves a high overall average reward and fairness index in comparison to the benchmarks. Concretely, the overall average reward of the proposal is the highest with 0.51, while the figure for the Origin-BBR, M-BBR, BBR+, BBR-Tsunami and Cubic are 0.43, 0.39, 0.45, 0.4 and 0.04, respectively. The overall average reward of our proposal shows an improvement of 18.6%, 30.76%, 13.3% and 27.5% in comparison with the figure for Origin-BBR and three other BBR's variants.

Despite the promising results, the novel troubleshooting framework has several limitations as follows.

– In this framework, we consider congestion as a use-case to generate anomalies and build troubleshooting datasets related to three root causes: link failure, switch failure and buffer overload. However, there are various kinds of anomalies (e.g. server disruption, etc.), as well as its root causes, are not yet taken into account in the framework.

– Encrypted traffic not only leads to a lack of information about application class but also results in many obstacles for network performance monitoring approaches (e.g. QoE estimation, etc.) and intrusion detection systems using a deep packet inspection (e.g. DDoS detection, botnet detection, etc.). However, we only take into account the application class as a use case in the novel troubleshooting framework.

– In the traffic classification approach, we use supervised learning algorithms to classify the network traffic into five applications, but it requires a large labeled dataset. Besides, building large labeled datasets is time consuming and costly. Consequently, building larger datasets with more kinds of applications will face many obstacles in the future.

– According to the related work, there are many remediation approaches in network troubleshooting such as routing, load balancing, etc. However, we take into account the routing aspect as an example in the temporary remediation module. The routing aspect can effectively reduce negative impacts of congestion in the network devices (e.g. router, switch, etc.), but it is ineffective with other anomalies (e.g. server disruption, switch configuration, etc.), which is not yet considered in the framework.

– Congestion can be caused by link failure, switch failure or buffer overload. In this book, buffer overload is taken into account, and we present a congestion control algorithm to solve it definitively. However, remediation approaches for other root causes need to be studied in the troubleshooting framework.

C.2. Prospects and future work

According to promising results, this book opens up several research directions.

– Firstly, troubleshooting datasets should include more data related to other root causes to facilitate network troubleshooting. Besides, a root cause analysis needs to be studied to identify the root causes in this case.

– The emergence of encrypted traffic causes limitations for network management and network troubleshooting regarding network performance monitoring approaches (e.g. QoE estimation, application identification, etc.) and intrusion detection system using a deep packet inspection (e.g. DDoS detection, botnet detection, etc.). In this book, we place a special focus on traffic classification solutions for identifying application classes in the context of encrypted traffic. Other aspects need to be thoroughly taken into account in the future.

– The traffic classification approach uses supervised ML algorithms to classify encrypted network traffic. It requires large labeled datasets, but capturing large labeled datasets is a costly and time consuming. Consequently, many studies focus on traffic classification approaches using semi-supervised ML algorithms. The objective is to pre-train a model on a large unlabeled dataset and transfer learned weights to a new model that is retrained with a small labeled dataset. Therefore, these approaches will be taken into account as a part of our future work.

– The application-aware segment routing mechanism presented in this book is implemented for certain applications such as video streaming, file transfer and VoIP. This mechanism selects an appropriate routing strategy corresponding to a specific application to meet strict service-level agreement requirements. The next step of this book is to extend to develop a more effective proof of concept in a product environment.

– Finally, we present an adaptive BBR algorithm for QUIC to adapt to the unexpected changes of network environments. The purpose is to change sending rates at sender sides adaptively in dynamic network. We will extend this work to select appropriate congestion control algorithms (e.g. QUIC Cubic, QUIC New Reno, QUIC BBR, etc.) across various network conditions. In a specific algorithm, we will implement RL algorithm to select appropriate sending rates to improve network performance in dynamic network.

References

Abdeljaouad, I., Rachidi, H., Fernandes, S., Karmouch, A. (2010). Performance analysis of modern TCP variants: A comparison of Cubic, Compound and New reno. *2010 25th Biennial Symposium on Communications*. IEEE, Kingston.

Abdelmoniem, A.M. and Bensaou, B. (2015). Incast-aware switch-assisted TCP congestion control for data centers. *Global Communications Conference (GLOBECOM)*. IEEE, San Diego.

Abdullah, Z.N., Ahmad, I., Hussain, I. (2018). Segment routing in software defined networks: A survey. *IEEE Communications Surveys & Tutorials*, 21(1), 464–486.

AbuHmed, T., Mohaisen, A., Nyang, D. (2008). A survey on deep packet inspection for intrusion detection systems. arXiv preprint. arXiv:0803.0037.

Adami, D., Antichi, G., Garroppo, R.G., Giordano, S., Moore, A.W. (2015). Towards an SDN network control application for differentiated traffic routing. *2015 IEEE International Conference on Communications (ICC)*. IEEE, London.

Ahmed, M., Mahmood, A.N., Hu, J. (2016). A survey of network anomaly detection techniques. *Journal of Network and Computer Applications*, 60, 19–31.

Ahmad, I., Basheri, M., Iqbal, M.J., Rahim, A. (2018). Performance comparison of support vector machine, random forest, and extreme learning machine for intrusion detection. *IEEE Access*, 6, 33789–33795.

Al-Naymat, G., Al-Kasassbeh, M., Al-Harwari, E. (2018). Using machine learning methods for detecting network anomalies within SNMP-MIB dataset. *International Journal of Wireless and Mobile Computing*, 15(1), 67–76.

Al-Tashi, Q., Rais, H.M., Abdulkadir, S.J., Mirjalili, S., Alhussian, H. (2020). A review of grey wolf optimizer-based feature selection methods for classification. In *Evolutionary Machine Learning Techniques*, Mirjalili, S., Faris, H., Aljarah, I. (eds). Springer, Singapore.

Aliyu, A.L., Bull, P., Abdallah, A. (2017). Performance implication and analysis of the OpenFlow SDN protocol. *2017 31st International Conference on Advanced Information Networking and Applications Workshops (WAINA)*. IEEE, Taipei.

Amaral, P., Dinis, J., Pinto, P., Bernardo, L., Tavares, J., Mamede, H.S. (2016). Machine learning in software defined networks: Data collection and traffic classification. *2016 IEEE 24th International Conference on Network Protocols (ICNP)*. IEEE, Singapore.

Amirthalingam, K. and Moorhead, R.J. (1995). SNMP – An overview of its merits and demerits. *Proceedings of the Twenty-Seventh Southeastern Symposium on System Theory*. IEEE, Starkville.

Amour, L., Sami, S., Hoceini, S., Mellouk, A. (2015a). Building a large dataset for model-based QoE prediction in the mobile environment. *Proceedings of the 18th ACM International Conference on Modeling, Analysis and Simulation of Wireless and Mobile Systems*, New York.

Amour, L., Sami, S., Hoceini, S., Mellouk, A. (2015b). An open source platform for perceived video quality evaluation. *Proceedings of the 11th ACM Symposium on QoS and Security for Wireless and Mobile Networks*, New York.

Amour, L., Tong, V., Souihi, S., Tran, H.A., Mellouk, A. (2019). Quality estimation framework for encrypted traffic (q2et). *2019 IEEE Global Communications Conference (GLOBECOM)*. IEEE, Waikoloa.

Amour, L., Souihi, S., Mellouk, A., Mushtaq, S. (2020). Q2abr: Qoe-aware adaptive video bit rate solution. *International Journal of Communication Systems*, 33(10), e4204.

Antonakakis, M., Perdisci, R., Nadji, Y., Vasiloglou, N., Abu-Nimeh, S., Lee, W., Dagon, D. (2012). From throw-away traffic to bots: Detecting the rise of DGA-based malware. *21st {USENIX} Security Symposium ({USENIX} Security 12)*. USENIX Association, Berkeley.

Atary, A. and Bremler-Barr, A. (2016). Efficient round-trip time monitoring in OpenFlow networks. *IEEE INFOCOM 2016 – The 35th Annual IEEE International Conference on Computer Communications*. IEEE, San Francisco.

Attarha, S., Hosseiny, K.H., Mirjalily, G., Mizanian, K. (2017). A load balanced congestion aware routing mechanism for software defined networks. *2017 Iranian Conference on Electrical Engineering (ICEE)*. IEEE, Tehran.

Azzouni, A., Boutaba, R., Trang, N.T.M., Pujolle, G. (2018). sOFTDP: Secure and efficient OpenFlow topology discovery protocol. *NOMS 2018-2018 IEEE/IFIP Network Operations and Management Symposium*. IEEE, Taipei.

Bahnasse, A., Louhab, F.E., Oulahyane, H.A., Talea, M., Bakali, A. (2018). Novel SDN architecture for smart MPLS traffic engineering-DiffServ aware management. *Future Generation Computer Systems*, 87, 115–126.

Bannour, F., Souihi, S., Mellouk, A. (2017). Distributed SDN control: Survey, taxonomy, and challenges. *IEEE Communications Surveys & Tutorials*, 20(1), 333–354.

Bannour, F., Souihi, S., Mellouk, A. (2020). Adaptive distributed SDN controllers: Application to content-centric delivery networks. *Future Generation Computer Systems*, 113, 78–93.

Barakabitze, A.A., Mkwawa, I.-H., Sun, L., Ifeachor, E. (2018). QualitySDN: Improving video quality using MPTCP and segment routing in SDN/NFV. *2018 4th IEEE Conference on Network Softwarization and Workshops (NetSoft)*. IEEE, Montreal.

Benayas, F., Carrera, A., Iglesias, C.A. (2018). Towards an autonomic Bayesian fault diagnosis service for SDN environments based on a big data infrastructure. *2018 Fifth International Conference on Software Defined Systems (SDS)*. IEEE, Barcelona.

Bennacer, L., Ciavaglia, L., Chibani, A., Amirat, Y., Mellouk, A. (2012). Optimization of fault diagnosis based on the combination of Bayesian networks and case-based reasoning. *2012 IEEE Network Operations and Management Symposium*. IEEE, Maui.

Benson, T., Akella, A., Maltz, D.A. (2010). Network traffic characteristics of data centers in the wild. *Proceedings of the 10th ACM SIGCOMM Conference on Internet Measurement*. ACM, New York.

Berde, P., Gerola, M., Hart, J., Higuchi, Y., Kobayashi, M., Koide, T., Lantz, B., O'Connor, B., Radoslavov, P., Snow, W., Parulkar, G. (2014). ONOS: Towards an open, distributed SDN OS. *Proceedings of the Third Workshop on Hot Topics in Software Defined Networking*, ACM, Chicago.

Berde, P., Gerola, M., Hart, J., Higuchi, Y., Kobayashi, M., Koide, T., Lantz, B., O'Connor, B., Radoslavov, P., Snow, W., Parulkar, G. (2021). Segment routing application designing [Online]. Available at: https://wiki.onosproject.org/display/ONOS/Segment+Routing+Application+Design.

Biere, A. (2008). Picosat essentials. *Journal on Satisfiability, Boolean Modeling and Computation*, 4, 75–97.

Bildea, A., Alphand, O., Rousseau, F., Duda, A. (2015). Link quality estimation with the Gilbert-Elliot model for wireless sensor networks. *2015 IEEE 26th Annual International Symposium on Personal, Indoor, and Mobile Radio Communications (PIMRC)*. IEEE, Hong Kong.

Böhmer, K. and Rinderle-Ma, S. (2020). Mining association rules for anomaly detection in dynamic process runtime behavior and explaining the root cause to users. *Information Systems*, 90, 101438.

Brian, O. and José, P.O. (2021). Mininet [Online]. Available at: http://mininet.org/.

Bu, K., Wen, X., Yang, B., Chen, Y., Li, L.E., Chen, X. (2016). Is every flow on the right track?: Inspect SDN forwarding with rulescope. *IEEE INFOCOM 2016 – The 35th Annual IEEE International Conference on Computer Communications*. IEEE, San Francisco.

Casas, P., D'Alconzo, A., Fiadino, P., Bär, A., Finamore, A., Zseby, T. (2014). When YouTube does not work – Analysis of QoE-relevant degradation in Google CDN traffic. *IEEE Transactions on Network and Service Management*, 11(4), 441–457.

Casas, P., Fiadino, P., Wassermann, S., Traverso, S., D'Alconzo, A., Tego, E., Matera, F., Mellia, M. (2016). Unveiling network and service performance degradation in the wild with mplane. *IEEE Communications Magazine*, 54(3), 71–79.

Cascone, C., Sanvito, D., Pollini, L., Capone, A., Sanso, B. (2017). Fast failure detection and recovery in SDN with stateful data plane. *International Journal of Network Management*, 27(2), e1957.

Cérin, C., Coti, C., Delort, P., Diaz, F., Gagnaire, M., Gaumer, Q., Guillaume, N., Lous, J., Lubiarz, S., Raffaelli, J. et al. (2013). Downtime statistics of current cloud solutions. *International Working Group on Cloud Computing Resiliency, Tech. Rep.*, 1, 2.

Chen, J., Chen, M., Wei, X., Chen, B. (2019). Matrix differential decomposition-based anomaly detection and localization in NFV networks. *IEEE Access*, 7, 29320–29331.

Cheng, L.-C., Wang, K., Hsu, Y.-H. (2015). Application-aware routing scheme for SDN-based cloud datacenters. *2015 Seventh International Conference on Ubiquitous and Future Networks (ICUFN)*. IEEE, Sapporo.

Cherrared, S., Imadali, S., Fabre, E., Gössler, G., Yahia, I.G.B. (2019). A survey of fault management in network virtualization environments: Challenges and solutions. *IEEE Transactions on Network and Service Management*, 16(4), 1537–1551.

Chollet, F. (2016a). Keras [Online]. Available at: https://github.com/fchollet/keras.

Chollet, F. (2016b). Loss function [Online]. Available at: https://keras.io/losses/.

Cid-Fuentes, J.A., Szabo, C., Falkner, K. (2018). Adaptive performance anomaly detection in distributed systems using online SVMs. *IEEE Transactions on Dependable and Secure Computing*, 17(5), 928–941.

Cisco (2018). Cisco and Vodafone showcase mobile transport networking advancements via segment routing at Mobile World Congress.

Cisco (2020). Cisco to optimize NTT DOCOMO's 5G mobile backhaul for simpler, more flexible and scalable network operation. Press Release.

Cisco (2021a). Cisco white paper report: Encrypted traffic analytics.

Cisco (2021b). Cisco visual networking index: Forecast and trends, 2017–2022. White Paper.

Clemente, L. (2021). A quic implementation in pure go [Online]. Available at: https://github.com/lucas-clemente/quic-go.

Cui, Y., Li, T., Liu, C., Wang, X., Kühlewind, M. (2017). Innovating transport with quic: Design approaches and research challenges. *IEEE Internet Computing*, 21(2), 72–76.

Custura, A., Secchi, R., Fairhurst, G. (2018). Exploring DSCP modification pathologies in the internet. *Computer Communications*, 127, 86–94.

Deri, L., Martinelli, M., Bujlow, T., Cardigliano, A. (2014). NDPI: Open-source high-speed deep packet inspection. *2014 International Wireless Communications and Mobile Computing Conference (IWCMC)*. IEEE, Nicosia.

Dholakiya, D., Kshirsagar, T., Nayak, A. (2021). Survey of mininet challenges, opportunities, and application in software-defined network (SDN). In *Information and Communication Technology for Intelligent Systems*, Senjyu, T., Mahalle, P.N., Perumal, T., Joshi, A. (eds). Springer, Singapore.

Dimopoulos, G., Leontiadis, I., Barlet-Ros, P., Papagiannaki, K., Steenkiste, P. (2015). Identifying the root cause of video streaming issues on mobile devices. *Proceedings of the 11th ACM Conference on Emerging Networking Experiments and Technologies*. ACM, New York.

Dmitri, T. and George, W. (2021). Lsquic library [Online]. Available at: https://github.com/litespeedtech/lsquic

Dong, M., Meng, T., Zarchy, D., Arslan, E., Gilad, Y., Godfrey, B., Schapira, M. (2018). Pcc vivace: Online-learning congestion control. *15th USENIX Symposium on Networked Systems Design and Implementation NSDI 18*. USENIX Association, Berkeley.

Dong, C., Zhang, C., Lu, Z., Liu, B., Jiang, B. (2020). Cetanalytics: Comprehensive effective traffic information analytics for encrypted traffic classification. *Computer Networks*, 176, 107258.

Du, M., Li, F., Zheng, G., Srikumar, V. (2017). Deeplog: Anomaly detection and diagnosis from system logs through deep learning. *Proceedings of the 2017 ACM SIGSAC Conference on Computer and Communications Security*. ACM, New York.

Eberendu, A.C. (2016). Unstructured data: An overview of the data of big data. *International Journal of Computer Trends and Technology*, 38(1), 46–50.

Fan, Z. and Liu, R. (2017). Investigation of machine learning based network traffic classification. *2017 International Symposium on Wireless Communication Systems (ISWCS)*. IEEE, Bologna.

Fathi-Kazerooni, S., Kaymak, Y., Rojas-Cessa, R. (2019). Tracking user application activity by using machine learning techniques on network traffic. *2019 International Conference on Artificial Intelligence in Information and Communication (ICAIIC)*. IEEE, Okinawa.

Felter, B. (2021). 7 of the most famous recent DDoS attacks. vXchnge [Online]. Available at: https://www.vxchnge.com/blog/recent-ddos-attacks-on-companies.

Fernandes, G., Rodrigues, J.J., Carvalho, L.F., Al-Muhtadi, J.F., Proença, M.L. (2019). A comprehensive survey on network anomaly detection. *Telecommunication Systems*, 70(3), 447–489.

Finsterbusch, M., Richter, C., Rocha, E., Muller, J.-A., Hanssgen, K. (2013). A survey of payload-based traffic classification approaches. *IEEE Communications Surveys & Tutorials*, 16(2), 1135–1156.

Floodlight (n.d.). Floodlight project [Online]. Available at: http://www.projectfloodlight.org

Fonseca, P.C. and Mota, E.S. (2017). A survey on fault management in software-defined networks. *IEEE Communications Surveys & Tutorials*, 19(4), 2284–2321.

Fortes, S., Barco, R., Aguilar-García, A., Muñoz, P. (2015). Contextualized indicators for online failure diagnosis in cellular networks. *Computer Networks*, 82, 96–113.

Frery, J., Habrard, A., Sebban, M., Caelen, O., He-Guelton, L. (2017). Efficient top rank optimization with gradient boosting for supervised anomaly detection. *Joint European Conference on Machine Learning and Knowledge Discovery in Databases*. Springer, Cham, Skopje.

Frost, D. and Bryant, S. (2011). Packet loss and delay measurement for MPLS networks. Technical Report, RFC 6374.

Gagnaire, M., Diaz, F., Coti, C., Cerin, C., Shiozaki, K., Xu, Y., Delort, P., Smets, J.-P., Le Lous, J., Lubiarz, S. et al. (2012). Downtime statistics of current cloud solutions. Technical report, International Working Group on Cloud Computing Resiliency.

Gales, M.J., Ragni, A., AlDamarki, H., Gautier, C. (2009). Support vector machines for noise robust ASR. *2009 IEEE Workshop on Automatic Speech Recognition & Understanding*. IEEE, Moreno.

Gao, S., Xiang, C., Sun, C., Qin, K., Lee, T.H. (2018). Efficient Boolean modeling of gene regulatory networks via random forest based feature selection and best-fit extension. *2018 IEEE 14th International Conference on Control and Automation (ICCA)*. IEEE, Anchorage.

Garcia, N., Alcaniz, T., González-Vidal, A., Bernabe, J.B., Rivera, D., Skarmeta, A. (2021). Distributed real-time SlowDoS attacks detection over encrypted traffic using artificial intelligence. *Journal of Network and Computer Applications*, 173, 102871.

Gholami, M. and Akbari, B. (2015). Congestion control in software defined data center networks through flow rerouting. *23rd Iranian Conference on Electrical Engineering (ICEE)*. IEEE, Tehran.

Github (2021). Github troubleshooting dataset [Online]. Available at: https://tinyurl.com/y23kgdro.

Goodfellow, I., Bengio, Y., Courville, A. (2016). *Deep Learning*. MIT Press, Cambridge, MA.

Grigoriou, E., Barakabitze, A.A., Atzori, L., Sun, L., Pilloni, V. (2017). An SDN-approach for QoE management of multimedia services using resource allocation. *2017 IEEE International Conference on Communications (ICC)*. IEEE, Paris.

Hadem, P., Saikia, D.K., Moulik, S. (2021). An SDN-based intrusion detection system using SVM with selective logging for IP traceback. *Computer Networks*, 191, 108015.

He, K. and Sun, J. (2015). Convolutional neural networks at constrained time cost. *Proceedings of the IEEE Conference on Computer Vision and Pattern Recognition*. IEEE, Boston.

He, Y.-L., Zhang, X.-L., Ao, W., Huang, J.Z. (2018). Determining the optimal temperature parameter for softmax function in reinforcement learning. *Applied Soft Computing*, 70, 80–85.

Heller, B., Scott, C., McKeown, N., Shenker, S., Wundsam, A., Zeng, H., Whitlock, S., Jeyakumar, V., Handigol, N., McCauley, J. et al. (2013). Leveraging SDN layering to systematically troubleshoot networks. *Proceedings of the Second ACM SIGCOMM Workshop on Hot Topics in Software Defined Networking*. ACM, New York.

Hochenbaum, J., Vallis, O.S., Kejariwal, A. (2017). Automatic anomaly detection in the cloud via statistical learning. *arXiv preprint. arXiv:1704.07706*.

Hock, M., Bless, R., Zitterbart, M. (2017). Experimental evaluation of BBR congestion control. *2017 IEEE 25th International Conference on Network Protocols (ICNP)*. IEEE, Toronto.

Hong, J., Park, S., Yoo, J.-H., Hong, J.W.-K. (2020). Machine learning based SLA-aware VNF anomaly detection for virtual network management. *2020 16th International Conference on Network and Service Management (CNSM)*. IEEE, Izmir.

Huitema, C. and Rescorla, E. (2018). Issues and requirements for SNI encryption in TLS. Document, Internet Engineering Task Force.

Hwang, J., Yoo, J., Lee, S.-H., Jin, H.-W. (2015). Scalable congestion control protocol based on SDN in data center networks. *2015 IEEE Global Communications Conference (GLOBECOM)*. IEEE, San Diego.

IANA (2021). Service name and transport protocol port number registry.

Ingre, B., Yadav, A., Soni, A.K. (2017). Decision tree based intrusion detection system for NSL-KDD dataset. *International Conference on Information and Communication Technology for Intelligent Systems*. Springer, India.

Jabbar, M., Aluvalu, R., Reddy, S.S.S. (2017). Intrusion detection system using Bayesian network and feature subset selection. *2017 IEEE International Conference on Computational Intelligence and Computing Research (ICCIC)*. IEEE, Coimbatore.

Jay, N., Rotman, N., Godfrey, B., Schapira, M., Tamar, A. (2019). A deep reinforcement learning perspective on internet congestion control. *International Conference on Machine Learning*. PMLR, Long Beach.

Jeong, S., Lee, D., Hyun, J., Li, J., Hong, J.W.-K. (2017). Application-aware traffic engineering in software-defined network. *2017 19th Asia-Pacific Network Operations and Management Symposium (APNOMS)*. IEEE, Seoul.

Jin, C., Yang, Z., Wang, Z., Jordan, M.I. (2020). Provably efficient reinforcement learning with linear function approximation. *Conference on Learning Theory*, PMLR, Long Beach.

Johnson, S.D., Flesher, J.W., Jehng, J.-C.J., Ferej, A. (1993). Enhancing electrical troubleshooting skills in a computer-coached practice environment. *Interactive Learning Environments*, 3(3), 199–214.

Jonassen, D.H. and Hung, W. (2006). Learning to troubleshoot: A new theory-based design architecture. *Educational Psychology Review*, 18(1), 77–114.

Jurkiewicz, P., Wójcik, R., Domżał, J., Kamisiński, A. (2020). Testing implementation of FAMTAR: Adaptive multipath routing. *Computer Communications*, 149, 300–311.

Kanagevlu, R. and Aung, K.M.M. (2015). SDN controlled local re-routing to reduce congestion in cloud data center. *2015 International Conference on Cloud Computing Research and Innovation (ICCCRI)*. IEEE, Singapore.

Kandula, S., Sengupta, S., Greenberg, A., Patel, P., Chaiken, R. (2009). The nature of data center traffic: Measurements & analysis. *Proceedings of the 9th ACM SIGCOMM Conference on Internet Measurement*. ACM, Chicago.

Karagiannis, T., Broido, A., Faloutsos, M., Claffy, K. (2004). Transport layer identification of p2p traffic. *Proceedings of the 4th ACM SIGCOMM Conference on Internet Measurement*. ACM, Chicago.

Katz, D. and Ward, D. (2010). Bidirectional forwarding detection (BFD). Technical report, Internet Engineering Task Force.

Kawasaki, J., Mouri, G., Suzuki, Y. (2020). Comparative analysis of network fault classification using machine learning. *NOMS 2020-2020 IEEE/IFIP Network Operations and Management Symposium*. IEEE, Budapest.

Kazemian, P., Varghese, G., McKeown, N. (2012). Header space analysis: Static checking for networks. *NSDI*, 12, 113–126.

Kelly, B. (2021). The percentage of encrypted flows. *No Jitter* [Online]. Available at: https://www.nojitter.com/cisco-malware-detection-what-communications-folks-need-know.

Khan, S., Gani, A., Wahab, A.W.A., Guizani, M., Khan, M.K. (2016). Topology discovery in software defined networks: Threats, taxonomy, and state-of-the-art. *IEEE Communications Surveys & Tutorials*, 19(1), 303–324.

Khraisat, A., Gondal, I., Vamplew, P. (2018). An anomaly intrusion detection system using c5 decision tree classifier. *Pacific-Asia Conference on Knowledge Discovery and Data Mining*. Springer, Singapore.

Kim, G., Kim, C., Jeong, J., Parker, M., Kim, J. (2016a). Contention-based congestion management in large-scale networks. *2016 49th Annual IEEE/ACM International Symposium on Microarchitecture (MICRO)*. IEEE, Taipei.

Kim, W., Li, J., Hong, J.W.-K., Suh, Y.-J. (2016b). OFMon: OpenFlow monitoring system in ONOS controllers. *2016 IEEE NetSoft Conference and Workshops (NetSoft)*. IEEE, Seoul.

Kishore, P. (2021). Servicenow troubleshooting dataset [Online]. Available at: https://github.com/vanvantong/Cognitive-Project/blob/master/latest_ticket_data.csv.

Koenig, S. and Simmons, R.G. (1993). Complexity analysis of real-time reinforcement learning. *Proceedings of the Eleventh National Conference on Artificial Intelligence, AAAI'93*. AAAI Press, Washington, DC.

Kosek, M., Shreedhar, T., Bajpai, V. (2021). Beyond QUIC v1: A first look at recent transport layer IETF standardization efforts. *IEEE Communications Magazine*, 59(4), 24–29.

Kühlewind, M., Trammell, B., Bühler, T., Fairhurst, G., Gurbani, V. (2018). Challenges in network management of encrypted traffic. *arXiv preprint. arXiv:1810.09272*.

Kukreja, N., Alvizu, R., Kos, A., Maier, G., Morro, R., Capello, A., Cavazzoni, C. (2016). Demonstration of SDN-based orchestration for multi-domain segment routing networks. *2016 18th International Conference on Transparent Optical Networks (ICTON)*. IEEE, Trento.

Kushwaha, A., Sharma, S., Bazard, N., Gumaste, A., Mukherjee, B. (2020). Design, analysis, and a terabit implementation of a source-routing-based SDN data plane. *IEEE Systems Journal*, 15(1), 56–67.

Kwon, D., Natarajan, K., Suh, S.C., Kim, H., Kim, J. (2018). An empirical study on network anomaly detection using convolutional neural networks. *2018 IEEE 38th International Conference on Distributed Computing Systems (ICDCS)*. IEEE, Vienna.

Langley, A., Riddoch, A., Wilk, A., Vicente, A., Krasic, C., Zhang, D., Yang, F., Kouranov, F., Swett, I., Iyengar, J. et al. (2017). The quic transport protocol: Design and internet-scale deployment. *Proceedings of the Conference of the ACM Special Interest Group on Data Communication*. ACM, Los Angeles.

Lara, A., Kolasani, A., Ramamurthy, B. (2013). Network innovation using OpenFlow: A survey. *IEEE Communications Surveys & Tutorials*, 16(1), 493–512.

Lee, M.-C. and Sheu, J.-P. (2016). An efficient routing algorithm based on segment routing in software-defined networking. *Computer Networks*, 103, 44–55.

Lee, S., Levanti, K., Kim, H.S. (2014). Network monitoring: Present and future. *Computer Networks*, 65, 84–98.

Lee, S., Kim, J., Shin, S., Porras, P., Yegneswaran, V. (2017). Athena: A framework for scalable anomaly detection in software-defined networks. *2017 47th Annual IEEE/IFIP International Conference on Dependable Systems and Networks (DSN)*. IEEE, Denver.

Li, G., Dong, M., Ota, K., Wu, J., Li, J., Ye, T. (2016). Deep packet inspection based application-aware traffic control for software defined networks. *Global Communications Conference (GLOBECOM)*. IEEE, Washington, DC.

Li, Y., Cai, Z.-P., Xu, H. (2018). LLMP: Exploiting LLDP for latency measurement in software-defined data center networks. *Journal of Computer Science and Technology*, 33(2), 277–285.

Li, Z., Peng, S., Voyer, D., Xie, C., Liu, P., Qin, Z., Ebisawa, K., Previdi, S., Guichard, J. (2020). Problem statement and use cases of application-aware networking (APN). Document, Internet Engineering Task Force.

Liao, L. and Leung, V.C. (2016). LLDP based link latency monitoring in software defined networks. *2016 12th International Conference on Network and Service Management (CNSM)*. IEEE, Montreal.

Lim, H.-K., Kim, J.-B., Kim, K., Hong, Y.-G., Han, Y.-H. (2019). Payload-based traffic classification using multi-layer LSTM in software defined networks. *Applied Sciences*, 9(12), 2550.

Lin, S.-C., Akyildiz, I.F., Wang, P., Luo, M. (2016). QoS-aware adaptive routing in multi-layer hierarchical software defined networks: A reinforcement learning approach. *2016 IEEE International Conference on Services Computing (SCC)*. IEEE, San Francisco.

Lin, H., Kuang, P., Shicheng, W., Ying, L., Xing, L., Shuping, P. (2019). Application-aware IPV6 networking. *Telecommunications Science*, 36(8), 36.

Lopez-Martin, M., Carro, B., Sanchez-Esguevillas, A., Lloret, J. (2017). Network traffic classifier with convolutional and recurrent neural networks for internet of things. *IEEE Access*, 5, 18042–18050.

Lopez-Millan, G., Marin-Lopez, R., Pereniguez-Garcia, F. (2019). Towards a standard SDN-based IPsec management framework. *Computer Standards & Interfaces*, 66, 103357.

Lotfollahi, M., Siavoshani, M.J., Zade, R.S.H., Saberian, M. (2020). Deep packet: A novel approach for encrypted traffic classification using deep learning. *Soft Computing*, 24(3), 1999–2012.

Lustig, T. (2021). Quic statistic [Online]. Available at: https://sniansfblog.org/the-potential-impact-of-quic-will-it-replace-tcp-ip/.

Maimó, L.F., Gómez, Á.L.P., Clemente, F.J.G., Pérez, M.G., Pérez, G.M. (2018). A self-adaptive deep learning-based system for anomaly detection in 5G networks. *IEEE Access*, 6, 7700–7712.

Mammeri, Z. (2019). Reinforcement learning based routing in networks: Review and classification of approaches. *IEEE Access*, 7, 55916–55950.

Marx, R., Herbots, J., Lamotte, W., Quax, P. (2020). Same standards, different decisions: A study of QUIC and HTTP/3 implementation diversity. *Proceedings of the Workshop on the Evolution, Performance, and Interoperability of QUIC*. ACM, New York.

Mdini, M. (2019). Anomaly detection and root cause diagnosis in cellular networks. PhD Thesis, Ecole nationale supérieure Mines-Télécom Atlantique Bretagne Pays de la Loire.

Moriarty, K. and Morton, A. (2018). Effects of pervasive encryption on operators [Online]. Available at: draft-mm-wg-effect-encrypt-25.

Morris, N.M. and Rouse, W.B. (1985). Review and evaluation of empirical research in troubleshooting. *Human Factors*, 27(5), 503–530.

Nagaraj, K., Killian, C., Neville, J. (2012). Structured comparative analysis of systems logs to diagnose performance problems. *9th USENIX Symposium on Networked Systems Design and Implementation (NSDI 12)*, USENIX Association, San Jose.

Nie, X., Zhao, Y., Li, Z., Chen, G., Sui, K., Zhang, J., Ye, Z., Pei, D. (2019). Dynamic TCP initial windows and congestion control schemes through reinforcement learning. *IEEE Journal on Selected Areas in Communications*, 37(6), 1231–1247.

Niranjan, A., Prakash, A., Veena, N., Geetha, M., Shenoy, P.D., Venugopal, K. (2017). EBJRV: An ensemble of bagging, J48 and random committee by voting for efficient classification of intrusions. *2017 IEEE International WIE Conference on Electrical and Computer Engineering (WIECON-ECE)*. IEEE, Dehradun.

OpenDayLight (2019). Opendaylight project [Online]. Available at: http://www.opendaylight.org/.

Orsolic, I. and Skorin-Kapov, L. (2020). A framework for in-network QoE monitoring of encrypted video streaming. *IEEE Access*, 8, 74691–74706.

Orsolic, I., Pevec, D., Suznjevic, M., Skorin-Kapov, L. (2017). A machine learning approach to classifying YouTube QoE based on encrypted network traffic. *Multimedia Tools and Applications*, 76(21), 22267–22301.

O'Shea, K. and Nash, R. (2015). An introduction to convolutional neural networks. *arXiv preprint. arXiv:1511.08458*.

Otoum, Y. and Nayak, A. (2021). AS-IDS: Anomaly and signature based IDS for the Internet of Things. *Journal of Network and Systems Management*, 29(3), 1–26.

Pakzad, F., Portmann, M., Tan, W.L., Indulska, J. (2014). Efficient topology discovery in software defined networks. *8th International Conference on Signal Processing and Communication Systems (ICSPCS)*. IEEE, Australia.

Patel, S., Shukla, Y., Kumar, N., Sharma, T., Singh, K. (2020). A comparative performance analysis of TCP congestion control algorithms: Newreno, westwood, veno, bic, and cubic. *6th International Conference on Signal Processing and Communication (ICSC)*. IEEE, Noida.

Pedregosa, F., Varoquaux, G., Gramfort, A., Michel, V., Thirion, B., Grisel, O., Blondel, M., Prettenhofer, P., Weiss, R., Dubourg, V. et al. (2011). Scikit-learn: Machine learning in python. *Journal of Machine Learning Research*, 12, 2825–2830.

Pedregosa, F., Varoquaux, G., Gramfort, A., Michel, V., Thirion, B., Grisel, O., Blondel, M., Prettenhofer, P., Weiss, R., Dubourg, V. et al. (2021a). Multi-layer perceptron [Online]. Available at: https://scikit-learn.org/stable/modules/generated/sklearn.neural_network.MLPClassifier.html.

Pedregosa, F., Varoquaux, G., Gramfort, A., Michel, V., Thirion, B., Grisel, O., Blondel, M., Prettenhofer, P., Weiss, R., Dubourg, V. et al. (2021b). Support vector machine [Online]. Available at: https://scikit-learn.org/stable/modules/generated/sklearn.svm.LinearSVC.html.

Peng, S., Mao, J., Hu, R., Li, Z. (2020). Demo abstract: Apn6: Application-aware IPv6 networking. *IEEE INFOCOM 2020-IEEE Conference on Computer Communications Workshops (INFOCOM WKSHPS)*. IEEE, Toronto.

Perešíni, P., Kuźniar, M., Kostić, D. (2015). Monocle: Dynamic, fine-grained data plane monitoring. *Proceedings of the 11th ACM Conference on Emerging Networking Experiments and Technologies*. ACM, New York.

Perez, R.S. (1991). *Toward a Unified Theory of Problem Solving: Views from the Content Domains*. Lawrence Erlbaum Associates, Inc, Hillsdale.

Phan, X.T. and Fukuda, K. (2017). SDN-Mon: Fine-grained traffic monitoring framework in software-defined networks. *Journal of Information Processing*, 25, 182–190.

Phanishayee, A., Krevat, E., Vasudevan, V., Andersen, D.G., Ganger, G.R., Gibson, G.A., Seshan, S. (2008). Measurement and analysis of TCP throughput collapse in cluster-based storage systems. *FAST*, 8, 1–14.

Pongsakorn, U., Watashiba, Y., Ichikawa, K., Date, S., Iida, H. (2017). Application-aware network: Network route management using SDN based on application characteristics. *CSI Transactions on ICT*, 5(4), 375–385.

Primartha, R. and Tama, B.A. (2017). Anomaly detection using random forest: A performance revisited. *2017 International Conference on Data and Software Engineering (ICoDSE)*. IEEE, Palembang.

Qiu, J., Du, Q., He, Y., Lin, Y., Zhu, J., Yin, K. (2018). Performance anomaly detection models of virtual machines for network function virtualization infrastructure with machine learning. *International Conference on Artificial Neural Networks*. Springer.

Rego, A., Sendra, S., Jimenez, J.M., Lloret, J. (2019). Dynamic metric OSPF-based routing protocol for software defined networks. *Cluster Computing*, 22(3), 705–720.

Rehman Javed, A., Jalil, Z., Atif Moqurrab, S., Abbas, S., Liu, X. (2020). Ensemble adaboost classifier for accurate and fast detection of botnet attacks in connected vehicles. *Transactions on Emerging Telecommunications Technologies*, 33(10), e4088.

Rescorla, E. and Dierks, T. (2018). The transport layer security (TLS) protocol version 1.3. Technical report, Internet Engineering Task Force.

Revina, A., Buza, K., Meister, V.G. (2020). IT ticket classification: The simpler, the better. *IEEE Access*, 8, 193380–193395.

Rezaei, S., Kroencke, B., Liu, X. (2019). Large-scale mobile app identification using deep learning. *IEEE Access*, 8, 348–362.

RF (2021). *shorturl.at/bquyW*.

Safaldin, M., Otair, M., Abualigah, L. (2021). Improved binary gray wolf optimizer and SVM for intrusion detection system in wireless sensor networks. *Journal of Ambient Intelligence and Humanized Computing*, 12(2), 1559–1576.

Salman, O., Elhajj, I.H., Chehab, A., Kayssi, A. (2018). A multi-level internet traffic classifier using deep learning. *2018 9th International Conference on the Network of the Future (NOF)*. IEEE, Poznan.

Schaafstal, A., Schraagen, J.M., van Berlo, M. (2000). Cognitive task analysis and innovation of training: The case of structured troubleshooting. *Human Factors*, 42(1), 75–86.

Schatz, R., Hoßfeld, T., Casas, P. (2012). Passive YouTube QoE monitoring for ISPS. *2012 Sixth International Conference on Innovative Mobile and Internet Services in Ubiquitous Computing*. IEEE, Palermo.

Selenium (2021). Webdriver [Online]. Available at: https://www.seleniumhq.org/projects/webdriver/

Shurman, M.M., Khrais, R.M., Yateem, A.A. (2019). IoT denial-of-service attack detection and prevention using hybrid IDS. *2019 International Arab Conference on Information Technology (ACIT)*. IEEE, Al Ain.

Singh, D. and Singh, B. (2020). Investigating the impact of data normalization on classification performance. *Applied Soft Computing*, 97, 105524.

Sinha, D., Haribabu, K., Balasubramaniam, S. (2015). Real-time monitoring of network latency in software defined networks. *2015 IEEE International Conference on Advanced Networks and Telecommuncations Systems (ANTS)*. IEEE, Kolkata.

Slavık, P. (1997). A tight analysis of the greedy algorithm for set cover. *Journal of Algorithms*, 25(2), 237–254.

Sminesh, C.N., Kanaga, E.G.M., Ranjitha, K. (2018). A proactive flow admission and re-routing scheme for load balancing and mitigation of congestion propagation in SDN data plane. *arXiv preprint. arXiv:1812.02474*.

Sokolova, M. and Lapalme, G. (2009). A systematic analysis of performance measures for classification tasks. *Information Processing & Management*, 45(4), 427–437.

Song, S., Lee, J., Son, K., Jung, H., Lee, J. (2016). A congestion avoidance algorithm in SDN environment. *2016 International Conference on Information Networking (ICOIN)*. IEEE, Kota Kinabalu.

Statista (2022). Data volume of global consumer IP traffic from 2017 to 2022. [Online]. Available at: https://www.statista.com/statistics/267202/global-data-volume-of-consumer-ip-traffic/.

Suárez-Varela, J. and Barlet-Ros, P. (2017). Towards a netflow implementation for OpenFlow software-defined networks. *2017 29th International Teletraffic Congress (ITC 29)*, vol. 1. IEEE, Genoa.

Tarnaras, G., Haleplidis, E., Denazis, S. (2015). SDN and forces based optimal network topology discovery. *Proceedings of the 2015 1st IEEE Conference on Network Softwarization (NetSoft)*. IEEE, London.

Taud, H. and Mas, J. (2018). Multilayer perceptron (MLP). In *Geomatic Approaches for Modeling Land Change Scenarios*, Olmedo, M.T.C., Paegelow, M., Mas, J.-F., Escobar, F. (eds). Springer, Cham.

Taylor, C.R., MacFarland, D.C., Smestad, D.R., Shue, C.A. (2016). Contextual, flow-based access control with scalable host-based SDN techniques. *IEEE INFOCOM 2016 – The 35th Annual IEEE International Conference on Computer Communications*. IEEE, San Francisco.

Team, O. (2021). Portstatistics API [Online]. Available at: http://api.onosproject.org/1.5.1/org/onosproject/net/device/PortStatistics.html.

Temlyakov, V. (2015). Greedy approximation in convex optimization. *Constructive Approximation*, 41(2), 269–296.

Thomas, V.P.S., Prasad, S.S., Kataoka, K. (2016). AMPF: Application-aware multipath packet forwarding using machine learning and SDN. *arXiv preprint arXiv:1606.05743*.

Tong, V. (2021a). Netflow-QUIC [Online]. Available at: https://drive.google.com/drive/folders/1cwHhzvaQbi-ap8yfrj2vHyPmUTQhaYOj?usp=sharing.

Tong, V. (2021b). Root cause analysis using machine learning [Online]. Available at: https://github.com/vanvantong/rca-ml.

Tong, V. (2021c). Service-centric segment routing using reinforcement learning [Online]. Available at: https://github.com/vanvantong/rl-sr.

Tong, V., Tran, H.A., Souihi, S., Mellouk, A. (2018). A novel QUIC traffic classifier based on convolutional neural networks. *2018 IEEE Global Communications Conference (GLOBECOM)*. IEEE, Abu Dhabi.

Tran, H.A., Hoceini, S., Mellouk, A., Perez, J., Zeadally, S. (2013). QoE-based server selection for content distribution networks. *IEEE Transactions on Computers*, 63(11), 2803–2815.

Tran, D., Mac, H., Tong, V., Tran, H.A., Nguyen, L.G. (2018). A LSTM based framework for handling multiclass imbalance in DGA botnet detection. *Neurocomputing*, 275, 2401–2413.

Tran, H.-A., Souihi, S., Tran, D., Mellouk, A. (2019). Mabrese: A new server selection method for smart SDN-based CDN architecture. *IEEE Communications Letters*, 23(6), 1012–1015.

Trestian, R., Muntean, G.-M., Katrinis, K. (2013). MiceTrap: Scalable traffic engineering of datacenter mice flows using OpenFlow. *2013 IFIP/IEEE International Symposium on Integrated Network Management (IM 2013)*. IEEE, Ghent.

Ujjan, R.M.A., Pervez, Z., Dahal, K., Bashir, A.K., Mumtaz, R., González, J. (2020). Towards sFlow and adaptive polling sampling for deep learning based DDoS detection in SDN. *Future Generation Computer Systems*, 111, 763–779.

Van, T., Tran, H.A., Souihi, S., Mellouk, A. (2018). Network troubleshooting: Survey, taxonomy and challenges. *2018 International Conference on Smart Communications in Network Technologies (SaCoNeT)*. IEEE, El Oued.

Vargas-Muñoz, M., Martínez-Peláez, R., Velarde-Alvarado, P., Moreno-García, E., Torres-Roman, D., Ceballos-Mejía, J. (2018). Classification of network anomalies in flow level network traffic using Bayesian networks. *2018 International Conference on Electronics, Communications and Computers (CONIELECOMP)*. IEEE, Cholula.

Velan, P., Čermák, M., Čeleda, P., Drašar, M. (2015). A survey of methods for encrypted traffic classification and analysis. *International Journal of Network Management*, 25(5), 355–374.

Ventre, P.L., Salsano, S., Polverini, M., Cianfrani, A., Abdelsalam, A., Filsfils, C., Camarillo, P., Clad, F. (2020). Segment routing: A comprehensive survey of research activities, standardization efforts and implementation results. *IEEE Communications Surveys & Tutorials*.

Wang, H., Gu, J., Wang, S. (2017a). An effective intrusion detection framework based on SVM with feature augmentation. *Knowledge-Based Systems*, 136, 130–139.

Wang, W., Zhu, M., Wang, J., Zeng, X., Yang, Z. (2017b). End-to-end encrypted traffic classification with one-dimensional convolution neural networks. *2017 IEEE International Conference on Intelligence and Security Informatics (ISI)*. IEEE, Beijing.

Wang, P., Ye, F., Chen, X., Qian, Y. (2018a). Datanet: Deep learning based encrypted network traffic classification in SDN home gateway. *IEEE Access*, 6, 55380–55391.

Wang, Y., Zhao, K., Li, W., Fraire, J., Sun, Z., Fang, Y. (2018b). Performance evaluation of QUIC with BBR in satellite internet. *2018 6th IEEE International Conference on Wireless for Space and Extreme Environments (WiSEE)*. IEEE, Huntsville.

Wang, J., Zheng, Y., Ni, Y., Xu, C., Qian, F., Li, W., Jiang, W., Cheng, Y., Cheng, Z., Li, Y. et al. (2019). An active-passive measurement study of TCP performance over LTE on high-speed rails. *The 25th Annual International Conference on Mobile Computing and Networking*, ACM, Los Cabos.

Wang, J., Tang, Y., He, S., Zhao, C., Sharma, P.K., Alfarraj, O., Tolba, A. (2020). Logevent2vec: Logevent-to-vector based anomaly detection for large-scale logs in Internet of Things. *Sensors*, 20(9), 2451.

Williams, N., Zander, S., Armitage, G. (2006). A preliminary performance comparison of five machine learning algorithms for practical IP traffic flow classification. *ACM SIGCOMM Computer Communication Review*, 36(5), 5–16.

Winstein, K. and Balakrishnan, H. (2013). TCP ex machina: Computer-generated congestion control. *ACM SIGCOMM Computer Communication Review*, 43(4), 123–134.

Woodbridge, J., Anderson, H.S., Ahuja, A., Grant, D. (2016). Predicting domain generation algorithms with long short-term memory networks. *arXiv preprint. arXiv:1611.00791*.

Wu, X., Wang, L., Bai, B., Qiao, J., Fei, A. (2020). HOPF bifurcation in DCTCP congestion control. *IEEE Communications Letters*, 24(7), 1424–1427.

Xu, H., Yan, L., Xing, H., Cui, Y., Li, S. (2017). Link failure detection in software defined networks: An active feedback mechanism. *Electronics Letters*, 53(11), 722–724.

Xu, Z., Tang, J., Yin, C., Wang, Y., Xue, G. (2019). Experience-driven congestion control: When multi-path TCP meets deep reinforcement learning. *IEEE Journal on Selected Areas in Communications*, 37(6), 1325–1336.

Yap, K.-K., Motiwala, M., Rahe, J., Padgett, S., Holliman, M., Baldus, G., Hines, M., Kim, T., Narayanan, A., Jain, A. et al. (2017). Taking the edge off with espresso: Scale, reliability and programmability for global internet peering. *Proceedings of the Conference of the ACM Special Interest Group on Data Communication*, ACM, Los Angeles.

Yu, X., Joshi, P., Xu, J., Jin, G., Zhang, H., Jiang, G. (2016). Cloudseer: Workflow monitoring of cloud infrastructures via interleaved logs. *ACM SIGARCH Computer Architecture News*, 44(2), 489–502.

Yu, Y., Li, X., Leng, X., Song, L., Bu, K., Chen, Y., Yang, J., Zhang, L., Cheng, K., Xiao, X. (2018). Fault management in software-defined networking: A survey. *IEEE Communications Surveys & Tutorials*, 21(1), 349–392.

Zeng, H., Kazemian, P., Varghese, G., McKeown, N. (2012a). Automatic test packet generation. *Proceedings of the 8th International Conference on Emerging Networking Experiments and Technologies*. ACM, Nice.

Zeng, H., Kazemian, P., Varghese, G., McKeown, N. (2012b). A survey on network troubleshooting. Technical report TR12-HPNG-061012, Stanford University.

Zhang, S. (2019). An evaluation of BBR and its variants. *arXiv preprint. arXiv:1909.03673*.

Zhang, T. and Mao, S. (2020). Machine learning for end-to-end congestion control. *IEEE Communications Magazine*, 58(6), 52–57.

Zhang, L., Cui, Y., Wang, M., Yang, Z., Jiang, Y. (2019). Machine learning for internet congestion control: Techniques and challenges. *IEEE Internet Computing*, 23(5), 59–64.

Zhao, Y., Zhang, P., Wang, Y., Jin, Y. (2018a). SDN-enabled rule verification on data plane. *IEEE Communications Letters*.

Zhao, Y., Zhang, P., Wang, Y., Jin, Y. (2018b). Troubleshooting data plane with rule verification in software-defined networks. *IEEE Transactions on Network and Service Management*, 15(1), 232–244.

Zhou, H., Li, Y., Yang, H., Jia, J., Li, W. (2018). Bigroots: An effective approach for root-cause analysis of stragglers in big data system. *IEEE Access*, 6, 41966–41977.

Zhou, K., Wang, W., Wu, C., Hu, T. (2020). Practical evaluation of encrypted traffic classification based on a combined method of entropy estimation and neural networks. *ETRI Journal*, 42(3), 311–323.

Zolotukhin, M. and Hämäläinen, T. (2018). Data stream clustering for application-layer DDoS detection in encrypted traffic. In *Cyber Security: Power and Technology*, Lehto, M. and Neittaanmäki, P. (eds). Springer, Cham.

Index

A, B, C

anomaly detection, 67–73, 75–79
application-aware segment routing, 81
BBR (bottleneck bandwidth and round-trip time), 121–133
buffer overload, 2, 7, 10, 23
congestion control, 107, 121–127, 129, 131–133
convolutional neural network, 63, 65

D, I, L

dataset, 37, 39–43
information theory, 68, 70
link failure, 2, 4–6, 10, 23

M, N, P

machine learning, 71
monitoring tools, 29, 32, 34
network troubleshooting, 1, 2, 9–13, 18, 22, 23
payload, 46, 47, 50–53, 55, 56, 64

Q, R

QUIC (Quick UDP Internet Connections), 10–16, 18–21, 45–49, 53, 54, 56, 59, 65
reinforcement learning, 83, 94, 105
root cause analysis, 107–112, 119
rule failure, 3, 4, 23
rule induction, 68, 70

S, T

SDN (software-defined networking), 25, 28–31, 37, 40, 43, 81–83, 85–88, 93, 97, 105, 107, 108, 111, 119
TLS (Transport Layer Security), 10–14, 16–21, 23
troubleshooting framework, 28, 29

Other titles from

iSTE

in

Networks and Telecommunications

2023

RAVALIMINOARIMALALASON Toky Basilide, RANDIMBINDRAINIBE Falimanana
Queues Applied to Telecoms: Courses and Exercises

2022

BANNOUR Fetia, SOUIHI Sami, MELLOUK Abdelhamid
Software-Defined Networking: Extending SDN Control to Large-Scale Networks
(New Generation Networks Set – Volume 2)

BENAROUS Leila, BITAM Salim, MELLOUK Abdelhamid
Security in Vehicular Networks: Focus on Location and Identity Privacy
(New Generation Networks Set – Volume 1)

2021

LAUNAY Frédéric
NG-RAN and 5G-NR: 5G Radio Access Network and Radio Interference

2020

PUJOLLE Guy
Software Networks: Virtualization, SDN, 5G and Security (2nd edition revised and updated)
(Advanced Network Set – Volume 1)

GONTRAND Christophe
Digital Communication Techniques

2019

LAUNEY Frédéric, PEREZ André
LTE Advanced Pro: Towards the 5G Mobile Network
Harmonic Concept and Applications

TOUNSI Wiem
Cyber-Vigilance and Digital Trust: Cyber Security in the Era of Cloud Computing and IoT

2018

ANDIA Gianfranco, DURO Yvan, TEDJINI Smail
Non-linearities in Passive RFID Systems: Third Harmonic Concept and Applications

BOUILLARD Anne, BOYER Marc, LE CORRONC Euriell
Deterministic Network Calculus: From Theory to Practical Implementation

LAUNAY Frédéric, PEREZ André
LTE Advanced Pro: Towards the 5G Mobile Network

PEREZ André
Wi-Fi Integration to the 4G Mobile Network

2017

BENSLAMA Malek, BENSLAMA Achour, ARIS Skander
Quantum Communications in New Telecommunications Systems

HILT Benoit, BERBINEAU Marion, VINEL Alexey, PIROVANO Alain
Networking Simulation for Intelligent Transportation Systems: High Mobile Wireless Nodes

LESAS Anne-Marie, MIRANDA Serge
The Art and Science of NFC Programming
(Intellectual Technologies Set – Volume 3)

2016

AL AGHA Khaldoun, PUJOLLE Guy, ALI-YAHIYA Tara
Mobile and Wireless Networks
(Advanced Network Set – Volume 2)

BATTU Daniel
Communication Networks Economy

BENSLAMA Malek, BATATIA Hadj, MESSAI Abderraouf
Transitions from Digital Communications to Quantum Communications: Concepts and Prospects

CHIASSERINI Carla Fabiana, GRIBAUDO Marco, MANINI Daniele
Analytical Modeling of Wireless Communication Systems
(Stochastic Models in Computer Science and Telecommunication Networks Set – Volume 1)

EL FALLAH SEGHROUCHNI Amal, ISHIKAWA Fuyuki, HÉRAULT Laurent, TOKUDA Hideyuki
Enablers for Smart Cities

PEREZ André
VoLTE and ViLTE

2015

BENSLAMA Malek, BATATIA Hadj, BOUCENNA Mohamed Lamine
Ad Hoc Networks Telecommunications and Game Theory

BENSLAMA Malek, KIAMOUCHE Wassila, BATATIA Hadj
Connections Management Strategies in Satellite Cellular Networks

BERTHOU Pascal, BAUDOIN Cédric, GAYRAUD Thierry, GINESTE Matthieu
Satellite and Terrestrial Hybrid Networks

CUADRA-SÁNCHEZ Antonio, ARACIL Javier
Traffic Anomaly Detection

LE RUYET Didier, PISCHELLA Mylène
Digital Communications 1: Source and Channel Coding

PEREZ André
LTE and LTE Advanced: 4G Network Radio Interface

PISCHELLA Mylène, LE RUYET Didier
Digital Communications 2: Digital Modulations

2014

ANJUM Bushra, PERROS Harry
Bandwidth Allocation for Video under Quality of Service Constraints

BATTU Daniel
New Telecom Networks: Enterprises and Security

BEN MAHMOUD Mohamed Slim, GUERBER Christophe, LARRIEU Nicolas, PIROVANO Alain, RADZIK José
Aeronautical Air−Ground Data Link Communications

BITAM Salim, MELLOUK Abdelhamid
Bio-inspired Routing Protocols for Vehicular Ad-Hoc Networks

CAMPISTA Miguel Elias Mitre, RUBINSTEIN Marcelo Gonçalves
Advanced Routing Protocols for Wireless Networks

CHETTO Maryline
Real-time Systems Scheduling 1: Fundamentals
Real-time Systems Scheduling 2: Focuses

EXPOSITO Ernesto, DIOP Codé
Smart SOA Platforms in Cloud Computing Architectures

MELLOUK Abdelhamid, CUADRA-SANCHEZ Antonio
Quality of Experience Engineering for Customer Added Value Services

OTEAFY Sharief M.A., HASSANEIN Hossam S.
Dynamic Wireless Sensor Networks

PEREZ André
Network Security

PERRET Etienne
Radio Frequency Identification and Sensors: From RFID to Chipless RFID

REMY Jean-Gabriel, LETAMENDIA Charlotte
LTE Standards
LTE Services

TANWIR Savera, PERROS Harry
VBR Video Traffic Models

VAN METER Rodney
Quantum Networking

XIONG Kaiqi
Resource Optimization and Security for Cloud Services

2013

ASSING Dominique, CALÉ Stéphane
Mobile Access Safety: Beyond BYOD

BEN MAHMOUD Mohamed Slim, LARRIEU Nicolas, PIROVANO Alain
Risk Propagation Assessment for Network Security: Application to Airport Communication Network Design

BERTIN Emmanuel, CRESPI Noël
Architecture and Governance for Communication Services

BEYLOT André-Luc, LABIOD Houda
Vehicular Networks: Models and Algorithms

BRITO Gabriel M., VELLOSO Pedro Braconnot, MORAES Igor M.
Information-Centric Networks: A New Paradigm for the Internet

DEUFF Dominique, COSQUER Mathilde
User-Centered Agile Method

DUARTE Otto Carlos, PUJOLLE Guy
Virtual Networks: Pluralistic Approach for the Next Generation of Internet

FOWLER Scott A., MELLOUK Abdelhamid, YAMADA Naomi
LTE-Advanced DRX Mechanism for Power Saving

JOBERT Sébastien *et al.*
Synchronous Ethernet and IEEE 1588 in Telecoms: Next Generation Synchronization Networks

MELLOUK Abdelhamid, HOCEINI Said, TRAN Hai Anh
Quality-of-Experience for Multimedia: Application to Content Delivery Network Architecture

NAIT-SIDI-MOH Ahmed, BAKHOUYA Mohamed, GABER Jaafar, WACK Maxime
Geopositioning and Mobility

PEREZ André
Voice over LTE: EPS and IMS Networks

2012

AL AGHA Khaldoun
Network Coding

BOUCHET Olivier
Wireless Optical Communications

DECREUSEFOND Laurent, MOYAL Pascal
Stochastic Modeling and Analysis of Telecoms Networks

DUFOUR Jean-Yves
Intelligent Video Surveillance Systems

EXPOSITO Ernesto
Advanced Transport Protocols: Designing the Next Generation

JUMIRA Oswald, ZEADALLY Sherali
Energy Efficiency in Wireless Networks

KRIEF Francine
Green Networking

PEREZ André
Mobile Networks Architecture

2011

BONALD Thomas, FEUILLET Mathieu
Network Performance Analysis

CARBOU Romain, DIAZ Michel, EXPOSITO Ernesto, ROMAN Rodrigo
Digital Home Networking

CHABANNE Hervé, URIEN Pascal, SUSINI Jean-Ferdinand
RFID and the Internet of Things

GARDUNO David, DIAZ Michel
Communicating Systems with UML 2: Modeling and Analysis of Network Protocols

LAHEURTE Jean-Marc
Compact Antennas for Wireless Communications and Terminals: Theory and Design

PALICOT Jacques
Radio Engineering: From Software Radio to Cognitive Radio

PEREZ André
IP, Ethernet and MPLS Networks: Resource and Fault Management

RÉMY Jean-Gabriel, LETAMENDIA Charlotte
Home Area Networks and IPTV

TOUTAIN Laurent, MINABURO Ana
Local Networks and the Internet: From Protocols to Interconnection

2010

CHAOUCHI Hakima
The Internet of Things

FRIKHA Mounir
Ad Hoc Networks: Routing, QoS and Optimization

KRIEF Francine
Communicating Embedded Systems / Network Applications

2009

CHAOUCHI Hakima, MAKNAVICIUS Maryline
Wireless and Mobile Network Security

VIVIER Emmanuelle
Radio Resources Management in WiMAX

2008

CHADUC Jean-Marc, POGOREL Gérard
The Radio Spectrum

GAÏTI Dominique
Autonomic Networks

LABIOD Houda
Wireless Ad Hoc and Sensor Networks

LECOY Pierre
Fiber-optic Communications

MELLOUK Abdelhamid
End-to-End Quality of Service Engineering in Next Generation Heterogeneous Networks

PAGANI Pascal *et al.*
Ultra-wideband Radio Propagation Channel

2007

BENSLIMANE Abderrahim
Multimedia Multicast on the Internet

PUJOLLE Guy
Management, Control and Evolution of IP Networks

SANCHEZ Javier, THIOUNE Mamadou
UMTS

VIVIER Guillaume
Reconfigurable Mobile Radio Systems

Printed and bound by CPI Group (UK) Ltd, Croydon, CR0 4YY
13/09/2026
14961042-0002